MATH

Grade 7

Thomas J. Richards
Mathematics Teacher
Lamar Junior-Senior High School
Lamar, Missouri

This book is dedicated to our children — Alyx, Nathan, Fred S., Dawn, Molly, Ellen, Rashaun, Brianna, Michele, Bradley, BriAnne, Kristie, Caroline, Dominic, Corey, Lindsey, Spencer, Morgan, Brooke, Cody, Sydney — and to all children who deserve a good education and who love to learn.

McGraw-Hill Consumer Products

McGraw-Hill
Consumer Products

A Division of The **McGraw-Hill** *Companies*

Copyright © 1997 McGraw-Hill Consumer Products.
Published by McGraw-Hill Learning Materials, an imprint of
McGraw-Hill Consumer Products.

Send all inquiries to:
McGraw-Hill Consumer Products
250 Old Wilson Bridge Road
Worthington OH 43085

ISBN 1-57768-117-7

4 5 6 7 8 9 10 POH 03 02 01 00 99

Table of Contents

Chapter 1
Operations Involving Whole Numbers

Chapter 2
Operations Involving Fractions

Chapter 3
Operations Involving Decimals

Chapter 4
Ratio and Proportion

Chapter 5
Decimals, Fractions, and Percent

Chapter 6
Percent

Chapter 7
Interest

Chapter 8
Metric Measurement

Chapter 9
Geometry

Chapter 10
Perimeter and Area

Chapter 11
Volume

Chapter 12
Statistics and Probability

Contents

MATHEMATICS Series
of Units

SPECTRUM MATHEMATICS is a non-graded, consumable series for students who need special help with the basic skills of computation and problem solving. This successful series emphasizes skill development and practice, without complex terminology or abstract symbolism. Because of the nature of the content and the students for whom the series is intended, readability has been carefully controlled to comply with the mathematics level of each book.

Features:

• A **Pre-Test** at the beginning of each chapter helps determine a student's understanding of the chapter content. The Pre-Test enables students and teachers to identify specific skills that need attention.

• **Developmental exercises** are provided at the top of the page when new skills are introduced. These exercises involve students in learning and serve as an aid for individualized instruction or independent study.

• **Abundant opportunities for practice** follow the developmental exercises.

• **Problem-solving pages** enable students to apply skills to realistic problems they will meet in everyday life.

• A **Test** at the end of each chapter gives students and teachers an opportunity to check understanding. A **Mid-Book Test**, covering Chapters 1–6, and a **Final Test**, covering all chapters, provide for further checks of understanding.

• A **Record of Test Scores** is provided on page xvi of this book so students can chart their progress as they complete each chapter test.

• **Answers** to all problems and test items are included at the back of the book.

This is the third edition of *SPECTRUM MATHEMATICS*. The basic books have remained the same. Some new, useful features have been added.

New Features:

• **Scope and Sequence Charts** for the entire *Spectrum Mathematics* series are included on pages iv–v.

• **Problem-Solving Strategy Lessons** are included on pages vii–xiv. These pages may be used at any time. The purpose is to provide students with various approaches to problem-solving.

• An **Assignment Record Sheet** is provided on page xv.

Problem-Solving Strategies

Sometimes you **use information in a table** to solve problems. When using a table, it is important to choose the needed information from the table and **ignore the extra information**.

Worker	Hourly Wage	Units Made	Vacation Days Used
Elvina	$8.23	234	9
Jimalla	$6.29	199	4
Candid	$7.46	232	0
Wayne	$8.14	219	8

Jimalla worked 35 hours last week. How much did she earn last week?

To solve the problem, you need to find Jimalla's hourly wage in the table, then decide what operation to use. The wage is $6.29. Multiply to solve the problem.

Jimalla earned $_____ last week.

Use the table above to solve each problem.

1. How many units did the four workers make in all?

What information will you use to solve the problem?

_____

What operation will you use to solve the problem? _____

The four workers produced _____ units in all.

2. Elvina has 15 days of vacation this year. How many more days does she have to take?

What information will you use to solve the problem?

_____

What operation will you use to solve the problem? _____

She still has _____ days of vacation to take.

3. Candid took 29 hours to make the units shown. How many units did she make in one hour?

What information will you use to solve the problem?

_____

What operation will you use to solve the problem? _____

She made _____ units in one hour.

1.

2.

3.

Perfect score: 9   My score: _____

Problem-Solving Strategies

Sometimes you can **make a table** to solve problems.

A truck goes 45 miles each hour. How long will it take to go 90 miles? 225 miles?

Number of Hours	1	2	3	4	5
Number of Miles	45	90	135	180	

1×45 2×45 3×45 4×45 5×45

Making a table is useful when you are asked more than one question, based on given information. Complete the table for 5 hours. Then use the table to answer each question.

It will take _____ hours to go 90 miles.

It will take _____ hours to go 225 miles.

Make a table for each problem. Then use the table to answer each question.

1. Beth made 200 units in 4 days. She made the same number of units each day. How many units did she make in 1 day? How long would it take her to make 250 units?

Number of Days	1	2	3	4		
Number of Units	50	100				

$200 \div 4$ 2×50 3×50

She made _____ units in 1 day.

It would take _____ days to make 250 units.

2. Six people are on a bus. At each of the next 6 stops, 3 more people get on. No one gets off. How many people are on the bus after the third stop? after the sixth stop? At which stop were there 18 people on the bus?

Number of Stops	1	2	3			
Number of People On	9	12				

$6 + 3$ $9 + 3$ $12 + 3$

_____ people are on after the third stop.

_____ people are on after the sixth stop.

Eighteen people are on after the _____ stop.

1.

2.

Perfect score: 18   My score: _____

x

Problem-Solving Strategies

Sometimes you can **make or use a drawing** to solve problems.

Mark has a rectangular yard that is 8 meters wide and 10 meters long. He plants trees along each side of the yard. There is a tree at each corner of the yard. Each tree is 2 meters from the next one. How many trees will be needed?

Mark will need _____ trees.

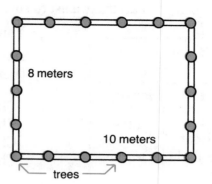

Make or use a drawing to solve each problem.

1. Each sprinkler waters a circular area with a 5-foot radius. Six sprinklers are placed in a straight line and spaced so the area watered overlaps by one foot. Find the length of the lawn watered with this placement.

_____ feet of lawn will be watered.

1.

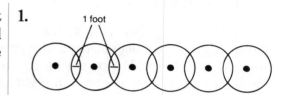

2. Jacob has an office that is a square with sides of 24 feet. The hallway outside his office is 5 feet wide and 20 feet long. Make a drawing to show the office and the hallways. How much carpeting will he need to carpet the office and the hallway?

_____ square feet of carpeting are needed.

2.

3. Lila has a garden that is 8 meters wide and 10 meters long. She plants 8-meter rows across the garden. She starts and finishes the rows on each edge of the garden. If the rows are .5 meters apart, how many rows will there be? Make a drawing, then answer the question.

There will be _____ rows.

3.

4. Four people are in line. Arnie is behind Barbara. Carol is in front of Arnie. Doris is in front of Carol. If Barbara is third in line, where is Carol? Make a drawing, then answer the question.

Carol is _____ in line.

4.

Perfect score: 4   My score: _____

Problem-Solving Strategies

Sometimes you can **make a list** to solve problems.

Choose 1 sandwich and 1 fruit	
Sandwiches	*Fruit*
Chicken	Apple
Roast Beef	Pear
Tuna Salad	Orange
Cheese	

You are to choose 1 sandwich and 1 fruit for lunch.

How many different combinations of sandwiches and fruit are there? Make a list to solve the problem.

Chicken, Apple	Cheese, Pear	Tuna Salad, Orange
Chicken, Pear	Cheese, Orange	Roast Beef, Apple
Chicken, Orange	Tuna Salad, Apple	Roast Beef, Pear
Cheese, Apple	Tuna Salad, Pear	Roast Beef, Orange

There are _____ different combinations.

Make a list to solve each problem.

Lists

1. Ken, Jolene, Ellyn, Bart, and Cal belong to a self-help group. Two of the people from the group are to attend a special meeting. Make a list to show the different combinations of two people who can attend the meeting. How many combinations are there?

1.

There are _____ combinations of 2 people.

2. Suppose that 3 people from the group above were to attend the meeting. Make a list to show the different combinations of three people. How many combinations are there?

2.

There are _____ combinations of 3 people.

3. Study the cartoon above. Make a list of why clear communications and written estimates are important. Include examples of possible problems and their solution.

_____

_____

_____

Perfect score: 3   My score: _____

Problem-Solving Strategies

NAME _____

Many times there is more than one way to solve a problem. You can **choose the strategy** that you like to use best.

Remember to follow the problem-solving plan.

You are in charge of committee finances. During one meeting the following transactions take place. Anna gave Ben $4. Ben gave Dan $5. Dan also was given $7 by Ellen. Ellen gave Anna and Ben $3 each. Dan gave Ellen $3. Each committee person started with just enough to pay each person they owed. How much did each person have at the start and at the end of the meeting?

┌─── **Plan** ───────────────┐
Read the problem.
Identify the question.
Identify the information you
   need to solve the problem.
Decide what strategy to use.
Use that strategy to find the
   answer.
Check that your answer makes
   sense.
└────────────────────────────┘

You could make a list or a table.

Anna		Ben		Dan		Ellen	
Gave	Got	Gave	Got	Gave	Got	Gave	Got
$4		$4			$5		
		$5			$7	$7	
	$3		$3			$3	
				$3		$3	
							$3
$	$	$	$	$	$	$	$

He started with ⤳ 🡕   🡔 He ended with

You could make a drawing.

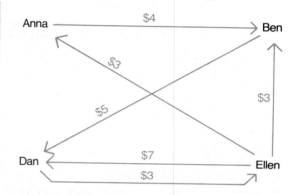

Now use the table or the drawing to solve the problem.

1. How much did Anna start with and end with?

Anna started with $_____ and ended with $_____.

2. How much did Ben start with and end with?

Ben started with $_____ and ended with $_____.

3. How much did Dan start with and end with?

Dan started with $_____ and ended with $_____.

4. How much did Ellen start with and end with?

Ellen started with $_____ and ended with $_____.

5. Another strategy might be to act out the problem. However, this would be hard to do. Why? (*Hint:* If you cannot think why, try acting out the problem.)

Perfect score: 9   My score: _____

Problem-Solving Strategies

Decide what strategy and/or what operation you need to use to solve each problem. Then solve each problem.

1. Twenty people are on a bus. Three people get off at each of the next four stops. No one gets on the bus. How many people are on the bus after the fourth stop?

_____ people are still on the bus.

1.

2. Five people are in line. Jack is directly in front of Misua. Misua is just in front of Berry. Anna is between Jack and Jane. What is Misua's position in the line?

Misua is _____ in line.

3. In **2**, who is at the front of the line?

_____ is in the front of the line.

2.-3.

4. Yolanda, Josh, Karyn, and Alan belong to a group. Two people are to attend a special meeting. How many different combinations of two people can attend the meeting?

There are _____ combinations.

5. Suppose 3 people are to attend the meeting in **4**. Then how many different combinations are there?

There are _____ combinations.

4.-5.

6. During a meeting the following transactions took place. Beverly gave Glenda $14. Dale gave Ginger $12. Glenda gave Ginger $10. Ginger gave Beverly and Dale $8 each. Find how much each person started with and ended with.

Beverly started with $_____ and ended with $_____.

Glenda started with $_____ and ended with $_____.

Dale started with $_____ and ended with $_____.

Ginger started with $_____ and ended with $_____.

6.

Perfect score: 13   My score: _____

Assignment Record Sheet

NAME _____

Pages Assigned	Date	Score	Pages Assigned	Date	Score	Pages Assigned	Date	Score

SPECTRUM MATHEMATICS

Record of Test Scores

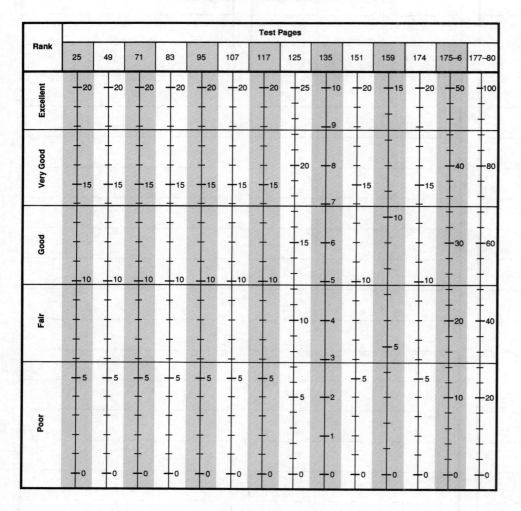

To record the score you receive on a TEST:

(1) Find the vertical scale below the page number of that TEST,

(2) on that vertical scale, draw a ● at the mark which represents your score.

For example, if your score for the TEST on page 25 is "My score: 13," draw a ● at the 13-mark on the first vertical scale. A score of 13 would show that your rank is "Good." You can check your progress from one test to the next by connecting the dots with a line segment.

Lesson 1   Addition

	Add ones.	Add tens.	Add hundreds.	Add thousands.	Add ten thousands.

$$
\begin{array}{r}
3\,3\,2\,3\,\overset{1}{4} \\
6\,0\,1\,2\,5 \\
+4\,3\,1\,4 \\
\hline
3
\end{array}
\qquad
\begin{array}{r}
3\,3\,2\,\overset{1}{3}\,4 \\
6\,0\,1\,2\,5 \\
+4\,3\,1\,4 \\
\hline
7\,3
\end{array}
\qquad
\begin{array}{r}
3\,3\,2\,\overset{1}{3}\,4 \\
6\,0\,1\,2\,5 \\
+4\,3\,1\,4 \\
\hline
6\,7\,3
\end{array}
\qquad
\begin{array}{r}
3\,3\,2\,\overset{1}{3}\,4 \\
6\,0\,1\,2\,5 \\
+4\,3\,1\,4 \\
\hline
7\,6\,7\,3
\end{array}
\qquad
\begin{array}{r}
3\,3\,2\,\overset{1}{3}\,4 \\
6\,0\,1\,2\,5 \\
+4\,3\,1\,4 \\
\hline
9\,7\,6\,7\,3
\end{array}
$$

$$4 + 5 + 4 = 13$$
$$13 = 10 + 3$$

Add.

	a	b	c	d	e
1.	34 +5	57 +412	317 +5142	1241 +17132	15434 +27112
2.	55 +5	68 +12	634 +49	258 +637	74057 +2639
3.	43 12 +14	516 127 +242	1618 2206 +154	30121 12314 +2121	61013 72342 +53624
4.	25 34 +13	162 270 +484	4180 1264 +3341	13136 21017 +32441	31453 10262 +14578
5.	72 33 51 +62	312 601 741 +113	1453 1622 2711 +3501	13164 21073 12241 +31120	13016 13147 24434 +36213
6.	48 25 34 56 +63	223 446 565 674 +193	6785 4215 6141 2573 +4122	30633 11710 20822 11201 +12411	29306 15447 44558 73775 +82862

Perfect score: 30   My score: _____

Problem Solving

Solve each problem.

1. There are 217 men and 232 women working at Woodson Electronics. How many people work there?

_____ people work at Woodson Electronics.

2. A store received a shipment of parts. One box contained 148 parts, another 132 parts, and the third 216 parts. How many parts were received in all?

The store received _____ parts in all.

3. Yesterday a warehouse received three shipments of merchandise. The weights of the shipments were 4,125 pounds, 912 pounds, and 2,436 pounds. What was the total weight of the shipments?

The total weight was _____ pounds.

4. During a 3-year period Mr. Sims drove his car the following distances: 9,546 miles, 8,742 miles, and 9,652 miles. How many miles did he drive his car during the 3 years?

He drove _____ miles.

5. Four new cars were sold last week. The prices were $6,422, $5,770, $7,282, and $4,864. What were the total sales for the week?

The total sales were $_____.

6. The population of Elmhurst is 43,526 and the population of Villa Park is 28,238. What is the combined population of the two towns?

The combined population is _____.

7. A creamery used the following amounts of milk: 89,421 kilograms, 76,125 kilograms, 48,129 kilograms, 69,743 kilograms, and 98,584 kilograms. What was the total amount used?

_____ kilograms were used.

1.	
2.	3.
4.	5.
6.	7.

Perfect score: 7   My score: _____

4

Lesson 2   Subtraction

NAME _____

Rename 9 tens and 4 as "8 tens and 14." Subtract ones.	Subtract tens.	Rename 8 thousands and 6 hundreds as "7 thousands and 16 hundreds." Subtract hundreds.	Subtract thousands.	Subtract ten thousands.
$\overset{8\ 14}{98\,69\,\cancel{4}}$ $-5\,67\,1\,6$ 8	$98\,6\overset{8\ 14}{\cancel{9}\cancel{4}}$ $-5\,67\,1\,6$ $7\,8$	$98\overset{7\ 16\ 8\ 14}{\cancel{6}\cancel{9}\cancel{4}}$ $-5\,6\,7\,1\,6$ $9\,7\,8$	$9\overset{7\ 16\ 8\ 14}{\cancel{8}\cancel{6}\cancel{9}\cancel{4}}$ $-5\,6\,7\,1\,6$ $1\,9\,7\,8$	$9\overset{7\ 16\ 8\ 14}{\cancel{8}\cancel{6}\cancel{9}\cancel{4}}$ $-5\,6\,7\,1\,6$ $4\,1\,9\,7\,8$

Subtract.

	a	b	c	d	e
1.	57 −4	447 −14	789 −191	5434 −675	57684 −4213
2.	68 −9	567 −19	847 −259	5678 −4132	78567 −40529
3.	23 −11	345 −94	804 −295	4567 −1418	51678 −10297
4.	76 −19	678 −89	4215 −104	4713 −2192	75159 −21642
5.	439 −7	804 −76	6758 −219	6507 −1706	75476 −19214
6.	675 −8	367 −151	4942 −261	7314 −2176	72143 −21674

Perfect score: 30   My score: _____

5

Problem Solving

Solve each problem.

1. Last Saturday the football team gained 87 yards rushing and 213 yards passing. How many more yards were gained passing than rushing?

_____ more yards were gained passing.

2. Anita's Scout troop sold 475 boxes of candles. Clara's troop sold 289 boxes of candles. How many more boxes did Anita's troop sell than Clara's troop?

Anita's troop sold _____ more boxes.

3. Mrs. Harris drove 912 kilometers in May. She drove 1,209 kilometers in June. How many more kilometers did she drive in June than in May?

She drove _____ more kilometers in June.

4. In Hillside 1,693 families take the evening paper and 1,275 take the morning paper. How many more families take the evening paper?

_____ more families take the evening paper.

5. In a contest 10,000 points are needed in order to win the grand prize. Milton has earned 7,975 points so far. How many more points does he need in order to win the grand prize?

He needs _____ more points.

6. Ten years ago the population of Rosedale was 9,675. Today the population is 12,245. How much has the population increased during the 10-year period?

The population has increased _____.

7. A truck and its cargo weigh 24,525 pounds. The truck weighs 12,750 pounds when empty. What is the weight of the cargo?

The cargo weighs _____ pounds.

1.	
2.	**3.**
4.	**5.**
6.	**7.**

Perfect score: 7   My score: _____

6

Lesson 3   Addition and Subtraction

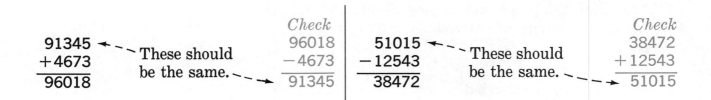

$$\begin{array}{r} 91345 \\ +4673 \\ \hline 96018 \end{array}$$

These should be the same.

Check
$$\begin{array}{r} 96018 \\ -4673 \\ \hline 91345 \end{array}$$

$$\begin{array}{r} 51015 \\ -12543 \\ \hline 38472 \end{array}$$

These should be the same.

Check
$$\begin{array}{r} 38472 \\ +12543 \\ \hline 51015 \end{array}$$

Add or subtract. Check each answer.

	a	b	c
1.	157 −36	2415 +332	63755 −1214
2.	125 +523	4769 −1251	75325 +11321
3.	679 −425	6958 +1035	84321 −29112
4.	785 +125	6007 −1539	37652 +92584
5.	708 −529	7395 +3764	30067 −14269

Perfect score: 15   My score: _____

7

Problem Solving

Solve each problem. Check each answer.

1. There were 345 people at the play Friday and 525 Saturday. How many people attended the play on those two days?

_____ people attended.

2. Henry threw a softball 132 feet and Walter threw it 119 feet. How much farther did Henry throw the softball than Walter?

Henry threw the softball _____ feet farther.

3. Jake delivered 1,225 circulars and Luke delivered 948. How many circulars did the two boys deliver?

The boys delivered _____ circulars.

4. The longest span of the Golden Gate Bridge is 1,280 meters. The longest span of the Brooklyn Bridge is 486 meters. What is the difference between those two distances?

The difference is _____ meters.

5. There are 3,529 books in the school library and 18,748 books in the city library. How many books are there in the two libraries?

There are _____ books in the two libraries.

6. Mt. McKinley is 20,320 feet high. Mt. Kennedy is 16,286 feet high. How much higher is Mt. McKinley than Mt. Kennedy?

Mt. McKinley is _____ feet higher.

7. The state of Illinois has an area of 56,400 square miles. The state of Wisconsin has an area of 56,154 square miles. What is the combined area of these two states?

The combined area is _____ square miles.

1.	
2.	3.
4.	5.
6.	7.

Perfect score: 7   My score: _____

8

Lesson 4 Addition and Subtraction

Add or subtract.

	a	b	c	d	e
1.	6 +43	67 +532	412 +5341	4725 +21063	56754 +43221
2.	59 −7	674 −52	8695 −364	67621 −5210	69548 −25037
3.	57 23 +11	634 26 +107	4125 516 +3243	51232 2426 +35319	12304 12130 +23149
4.	63 −17	876 −129	7654 −4126	54983 −1276	67894 −41327
5.	20 67 51 +21	120 63 281 +454	224 4052 3374 +126	2113 63012 11475 +2381	10168 54321 23146 +11287
6.	87 −49	507 −169	3769 −1595	63758 −1297	67885 −24389
7.	68 54 35 76 +46	724 156 345 654 +107	4124 124 5674 1176 +6782	3124 14765 23415 36767 +7412	54321 15247 61232 41234 +31259
8.	90 −47	700 −483	3007 −1249	50075 −1346	76548 −27879

Perfect score: 40   My score: _____

9

Problem Solving

Solve each problem.

1. Today's high temperature was 32 degrees Celsius. The low temperature was 19 degrees. What is the difference between those temperatures?

The difference is _____ degrees.

2. Ira worked 38 problems. Elise worked 47. Mabelle worked 63. How many problems did all three work?

They worked _____ problems.

3. The Poes are on a 2,027-kilometer trip. The first day they went 560 kilometers. How much farther do they still have to go?

They have to go _____ kilometers farther.

4. Mary and Martha collect stamps. Mary has 1,237 stamps, and Martha has 974 stamps. How many stamps do the two girls have?

The two girls have _____ stamps.

5. In problem 4, how many more stamps does Mary have than Martha?

Mary has _____ more stamps.

6. Of the 87,725 residents of Page County, 37,968 are registered voters. How many residents are not registered voters?

_____ are not registered voters.

7. What is the combined area of the Great Lakes?

The area is _____ square miles.

lake	area in square miles
Huron	23,010
Ontario	7,540
Michigan	22,400
Erie	9,940
Superior	31,820

1.	
2.	**3.**
4.	**5.**
6.	**7.**

Perfect score: 7   My score: _____

10

Lesson 6 Rounding Numbers

Study how to round a number to the nearest **ten**.

| Look at the *ones digit.* | 4 6 2 less than 5 | 4 6 5 equal to 5 | 4 6 8 more than 5 |

Round down to 460.　　　Round up to 470.　　　Round up to 470.

Study how to round a number to the nearest **hundred**.

| Look at the *tens digit.* | 8 2 1 9 less than 5 | 8 2 5 9 equal to 5 | 8 2 7 9 more than 5 |

Round down to 8200.　　　Round up to 8300.　　　Round up to 8300.

Study how to round a number to the nearest **thousand**.

| Look at the *hundreds digit.* | 6 0 3 5 less than 5 | 6 5 3 5 equal to 5 | 6 6 3 5 more than 5 |

Round down to 6000.　　　Round up to 7000.　　　Round up to 7000.

Round each number to the nearest ten.

	a	*b*	*c*	*d*
1.	48 _____	62 _____	51 _____	45 _____
2.	376 _____	429 _____	552 _____	245 _____

Round each number to the nearest hundred.

	a	*b*	*c*	*d*
3.	435 _____	681 _____	548 _____	957 _____
4.	8681 _____	3550 _____	3349 _____	2060 _____

Round each number to the nearest thousand.

	a	*b*	*c*	*d*
5.	4643 _____	8354 _____	6750 _____	5500 _____
6.	6455 _____	7500 _____	4271 _____	9814 _____

Perfect score: 24　　My score: _____

Lesson 7   Estimating Products

NAME _____

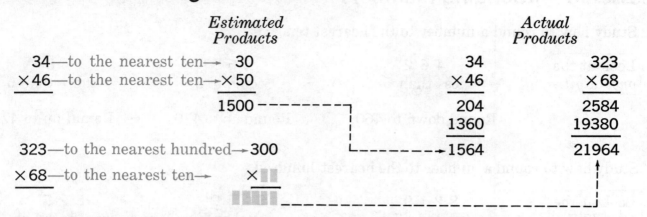

	Estimated Products			Actual Products	
34—to the nearest ten→	30		34		323
×46—to the nearest ten→	×50		×46		×68
	1500		204		2584
			1360		19380
323—to the nearest hundred→	300		1564		21964
×68—to the nearest ten→	×				

Write the estimated product on each blank. Then find each product.

	a	b	c
1.	96 ×62	63 ×88	91 ×81
2.	52 ×35	65 ×37	75 ×45
3.	498 ×98	801 ×82	502 ×52
4.	597 ×204	403 ×496	605 ×305

Perfect score: 24   My score: _____

14

Lesson 1   Multiplication

Multiply the numerators.

$$\frac{1}{2} \times \frac{3}{5} \times \frac{3}{4} = \frac{1 \times 3 \times 3}{2 \times 5 \times 4} = \frac{9}{40}$$

Multiply the denominators.

$$\frac{1}{2} \times \frac{3}{4} = \frac{1 \times 3}{2 \times 4} = \underline{\quad\quad} \qquad\qquad \frac{4}{5} \times \frac{2}{3} \times \frac{1}{3} = \frac{\quad \times \quad \times \quad}{\quad \times \quad \times \quad} = \underline{\quad\quad}$$

Multiply.

	a	b	c	d
1.	$\frac{3}{4} \times \frac{1}{5}$	$\frac{2}{3} \times \frac{4}{5}$	$\frac{7}{8} \times \frac{5}{6} \times \frac{1}{2}$	$\frac{3}{5} \times \frac{2}{7} \times \frac{1}{5}$
2.	$\frac{1}{2} \times \frac{1}{3}$	$\frac{4}{5} \times \frac{2}{7}$	$\frac{3}{8} \times \frac{3}{5} \times \frac{3}{4}$	$\frac{1}{2} \times \frac{1}{4} \times \frac{1}{3}$
3.	$\frac{5}{8} \times \frac{5}{8}$	$\frac{6}{7} \times \frac{3}{5}$	$\frac{2}{3} \times \frac{1}{5} \times \frac{1}{7}$	$\frac{3}{7} \times \frac{1}{5} \times \frac{3}{4}$
4.	$\frac{3}{8} \times \frac{5}{7}$	$\frac{5}{7} \times \frac{3}{8}$	$\frac{4}{5} \times \frac{4}{5} \times \frac{4}{5}$	$\frac{2}{3} \times \frac{2}{3} \times \frac{2}{3}$
5.	$\frac{3}{5} \times \frac{2}{5}$	$\frac{7}{8} \times \frac{3}{10}$	$\frac{5}{6} \times \frac{5}{9} \times \frac{1}{2}$	$\frac{3}{5} \times \frac{2}{5} \times \frac{4}{5}$
6.	$\frac{3}{5} \times \frac{1}{2}$	$\frac{5}{8} \times \frac{7}{9}$	$\frac{3}{8} \times \frac{5}{7} \times \frac{3}{4}$	$\frac{2}{3} \times \frac{4}{5} \times \frac{1}{7}$

Perfect score: 24   My score: _____

Lesson 2 Greatest Common Factor

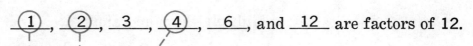

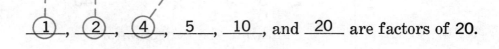

1, 2, and 4 are **common factors** of 12 and 20.
4 is the **greatest common factor** of 12 and 20.

List the factors of each number named below. Then list the common factors and the greatest common factor of each pair of numbers.

		factors	*common factor(s)*	*greatest common factor*
1.	4	_____		
	6	_____	_____	_____
2.	10	_____		
	12	_____	_____	_____
3.	16	_____		
	24	_____	_____	_____
4.	9	_____		
	16	_____	_____	_____
5.	18	_____		
	20	_____	_____	_____
6.	25	_____		
	24	_____	_____	_____
7.	48	_____		
	36	_____	_____	_____

Perfect score: 28   My score: _____

28

Lesson 3   Renaming in Simplest Form

A fraction is in simplest form when the greatest common factor of the numerator and denominator is 1.

Divide 8 and 12 by their greatest common factor.

$$\frac{8}{12} = \frac{8 \div 4}{12 \div 4} = \frac{2}{3}$$

The simplest form for $\frac{8}{12}$ is $\frac{2}{3}$.

A mixed numeral is in simplest form when its fraction is in simplest form and names a number between 0 and 1.

Divide 10 and 15 by their greatest common factor.

$$8\frac{10}{15} = 8\frac{10 \div 5}{15 \div 5} = 8\frac{2}{3}$$

The simplest form for $8\frac{10}{15}$ is _____.

Write each of the following in simplest form.

	a	*b*	*c*
1.	$\frac{6}{15}$	$\frac{12}{20}$	$\frac{21}{30}$
2.	$\frac{16}{36}$	$\frac{40}{42}$	$\frac{18}{36}$
3.	$2\frac{9}{15}$	$6\frac{18}{30}$	$8\frac{36}{54}$
4.	$\frac{10}{35}$	$3\frac{4}{18}$	$7\frac{18}{24}$
5.	$5\frac{18}{45}$	$\frac{30}{105}$	$\frac{36}{60}$

Perfect score: 15   My score:_____

Lesson 4 Expressing Products in Simplest Form

Study the two ways $\frac{6}{7} \times \frac{3}{4}$ is found in simplest form.

$$\frac{6}{7} \times \frac{3}{4} = \frac{6 \times 3}{7 \times 4}$$
$$= \frac{18}{28}$$
$$= \frac{18 \div 2}{28 \div 2}$$
$$= \underline{\quad\quad}$$

$$\frac{6}{7} \times \frac{3}{4} = \frac{\overset{3}{6} \times 3}{7 \times \underset{2}{4}}$$
$$= \frac{3 \times 3}{7 \times 2}$$
$$= \underline{\quad\quad}$$

Divide 6 (in the numerator) and 4 (in the denominator) by their greatest common factor, 2.

Write each answer in simplest form.

	a	b	c	d
1.	$\frac{4}{5} \times \frac{1}{2}$	$\frac{3}{8} \times \frac{2}{3}$	$\frac{3}{5} \times \frac{4}{9}$	$\frac{8}{9} \times \frac{7}{10}$
2.	$\frac{2}{3} \times \frac{5}{6}$	$\frac{7}{10} \times \frac{8}{9}$	$\frac{10}{11} \times \frac{7}{12}$	$\frac{5}{7} \times \frac{3}{10}$
3.	$\frac{7}{9} \times \frac{6}{11}$	$\frac{12}{13} \times \frac{3}{4}$	$\frac{10}{11} \times \frac{7}{15}$	$\frac{8}{9} \times \frac{5}{12}$
4.	$\frac{5}{8} \times \frac{2}{5}$	$\frac{9}{16} \times \frac{1}{6}$	$\frac{1}{2} \times \frac{8}{9}$	$\frac{4}{7} \times \frac{14}{15}$
5.	$\frac{3}{8} \times \frac{4}{5}$	$\frac{8}{9} \times \frac{6}{7}$	$\frac{4}{7} \times \frac{5}{6}$	$\frac{4}{15} \times \frac{12}{13}$

Perfect score: 20   My score: _____

Lesson 5   Products in Simplest Form

Study how $\frac{5}{6} \times \frac{9}{10} \times \frac{1}{7}$ is found in simplest form.

$$\frac{\overset{1}{\cancel{5}} \times 9 \times 1}{6 \times \underset{2}{\cancel{10}} \times 7}$$

Both numerator and denominator have been divided

by_____.

$$\frac{\overset{1}{\cancel{5}} \times \overset{3}{\cancel{9}} \times 1}{\underset{2}{\cancel{6}} \times \underset{2}{\cancel{10}} \times 7}$$

Both numerator and denominator have been divided

by_____.

$$\frac{\overset{1}{\cancel{5}} \times \overset{3}{\cancel{9}} \times 1}{\underset{2}{\cancel{6}} \times \underset{2}{\cancel{10}} \times 7} = \frac{1 \times 3 \times 1}{2 \times 2 \times 7}$$

$$= \frac{3}{28}$$

$$\frac{5}{6} \times \frac{9}{10} \times \frac{1}{7} = \text{\_\_\_\_\_}$$

Write each answer in simplest form.

	a	b	c	d
1.	$\frac{3}{5} \times \frac{5}{6}$	$\frac{4}{9} \times \frac{3}{8}$	$\frac{7}{8} \times \frac{6}{7}$	$\frac{2}{3} \times \frac{3}{4} \times \frac{1}{5}$
2.	$\frac{9}{10} \times \frac{5}{6}$	$\frac{5}{8} \times \frac{4}{15}$	$\frac{10}{11} \times \frac{11}{12}$	$\frac{4}{5} \times \frac{3}{4} \times \frac{5}{6}$
3.	$\frac{7}{9} \times \frac{9}{14}$	$\frac{4}{5} \times \frac{5}{12}$	$\frac{3}{4} \times \frac{8}{9}$	$\frac{4}{7} \times \frac{2}{3} \times \frac{7}{8}$
4.	$\frac{4}{9} \times \frac{9}{10}$	$\frac{2}{3} \times \frac{9}{10}$	$\frac{3}{10} \times \frac{5}{6}$	$\frac{2}{5} \times \frac{3}{4} \times \frac{5}{6}$
5.	$\frac{5}{9} \times \frac{9}{10}$	$\frac{7}{8} \times \frac{2}{7}$	$\frac{3}{10} \times \frac{5}{9}$	$\frac{4}{7} \times \frac{7}{8} \times \frac{2}{3}$

Perfect score: 20   My score: _____

Lesson 6   Renaming Numbers

Change $3\frac{2}{5}$ to a fraction.

$$3\frac{2}{5}=\frac{(5\times 3)+2}{5}$$

$$=\frac{15+2}{5}$$

$$=\frac{17}{5}$$

Change $\frac{22}{4}$ to a mixed numeral.

$\frac{22}{4}$ means $22\div 4$ or $4\overline{)22}$.

$$\begin{array}{r} 5\frac{2}{4} \\ 4\overline{)22} \\ \underline{20} \\ 2 \end{array} \quad 2\div 4=\frac{2}{4}$$

$$\frac{22}{4}=5\frac{2}{4} \text{ or } 5\frac{1}{2}$$

You can think of every whole number as a fraction with a denominator of 1.

$2=\frac{2}{1}$ $5=\frac{5}{1}$ $21=\frac{21}{1}$

Change each of the following to a fraction.

	a	b	c	d	e
1.	$1\frac{7}{10}$	$2\frac{1}{2}$	$4\frac{3}{5}$	$1\frac{3}{4}$	8
2.	$2\frac{3}{4}$	$3\frac{5}{6}$	$5\frac{1}{3}$	12	$6\frac{7}{8}$

Change each of the following to a mixed numeral in simplest form.

	a	b	c	d	e
3.	$\frac{9}{4}$	$\frac{21}{5}$	$\frac{9}{6}$	$\frac{18}{8}$	$2\frac{9}{5}$
4.	$\frac{43}{8}$	$\frac{64}{10}$	$\frac{22}{4}$	$\frac{16}{12}$	$5\frac{15}{10}$

Perfect score: 20   My score: _____

Lesson 7   Multiplication

NAME _____

$6 \times 5\frac{3}{4} = \frac{6}{1} \times \frac{23}{4}$ Rename the numbers as fractions.

$= \frac{\overset{3}{\cancel{6}} \times 23}{1 \times \underset{2}{\cancel{4}}}$ Divide the numerator and the denominator by common factors.

$= \frac{69}{2}$ Multiply.

$= 34\frac{1}{2}$ Write the product as a mixed numeral in simplest form.

$2\frac{2}{3} \times 1\frac{1}{2} \times \frac{4}{5} = \frac{8}{3} \times \frac{3}{2} \times \frac{4}{5}$

$= \frac{\overset{4}{\cancel{8}} \times \overset{1}{\cancel{3}} \times 4}{\underset{1}{\cancel{3}} \times \underset{1}{\cancel{2}} \times 5}$

$= \frac{4 \times 1 \times 4}{1 \times 1 \times 5}$

$= \frac{16}{5} \text{ or } 3\frac{1}{5}$

Write each answer in simplest form.

	a	*b*	*c*
1.	$8 \times 2\frac{5}{6}$	$4\frac{2}{3} \times 9$	$3\frac{1}{6} \times 2 \times 9$
2.	$1\frac{2}{3} \times 1\frac{1}{5}$	$1\frac{2}{7} \times 2\frac{1}{3}$	$1\frac{1}{3} \times 1\frac{1}{8} \times 1\frac{2}{3}$
3.	$1\frac{1}{9} \times \frac{3}{8}$	$\frac{5}{6} \times 1\frac{1}{8}$	$\frac{4}{5} \times 3\frac{1}{2} \times 2\frac{1}{2}$
4.	$3\frac{1}{3} \times 1\frac{1}{5}$	$2\frac{2}{5} \times \frac{5}{8}$	$2\frac{1}{3} \times \frac{3}{7} \times \frac{1}{2}$

Perfect score: 12   My score: _____

Problem Solving

Solve. Write each answer in simplest form.

1. Ava has $1\frac{1}{2}$ sacks of flour. Each sack weighs 5 pounds. How many pounds of flour does Ava have?

Ava has _____ pounds of flour.

2. Mr. DiMaggio bought $1\frac{3}{4}$ pounds of nuts. Two thirds of this amount was used at a dinner party. How many pounds of nuts were used?

_____ pounds of nuts were used.

3. An engine uses $2\frac{3}{4}$ gallons of fuel each hour. At that rate how many gallons of fuel would the engine use in $3\frac{2}{5}$ hours?

It would use _____ gallons.

4. In problem **3**, how many gallons of fuel would the engine use in 8 hours?

It would use _____ gallons.

5. Each inch on a map represents 18 miles on the ground. How many miles are represented by $2\frac{1}{4}$ inches on the map?

_____ miles are represented.

6. A carpenter has 12 boards each $5\frac{5}{6}$ feet long. What is the combined length of all 12 boards?

The combined length is _____ feet.

7. It takes 36 nails to weigh a pound. How many nails would it take to weigh $7\frac{2}{3}$ pounds?

There are _____ nails in $7\frac{2}{3}$ pounds.

8. Alva has $3\frac{4}{5}$ pounds of meat. John has $3\frac{2}{3}$ times as much meat as Alva. How many pounds of meat does John have?

John has _____ pounds of meat.

1.
2.
3.
4.
5.
6.
7.
8.

Perfect score: 8   My score: _____

Lesson 8   Reciprocals

If the product of two numbers is **1**, the numbers are **reciprocals** of each other.

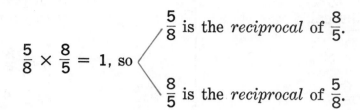

$$\frac{5}{8} \times \frac{8}{5} = 1, \text{ so}$$

$\frac{5}{8}$ is the *reciprocal* of $\frac{8}{5}$.

$\frac{8}{5}$ is the *reciprocal* of $\frac{5}{8}$.

reciprocals
$$\frac{4}{5} \times \frac{5}{4} = \frac{20}{20} \text{ or } 1$$

reciprocals
$$2 \times \frac{1}{2} = \frac{2}{2} \text{ or } 1$$

reciprocals
$$2\frac{1}{3} \times \frac{3}{7} = \frac{7}{3} \times \frac{3}{7} = \frac{21}{21} \text{ or } 1$$

Tell whether the fractions in each pair are reciprocals.
Write *Yes* or *No*.

	a	*b*	*c*
1.	$\frac{3}{4}$ and $\frac{4}{3}$ _____	$\frac{1}{5}$ and $\frac{5}{1}$ _____	$2\frac{3}{4}$ and $2\frac{4}{3}$ _____
2.	7 and $\frac{7}{1}$ _____	$3\frac{2}{5}$ and $\frac{5}{17}$ _____	$\frac{9}{10}$ and $\frac{9}{10}$ _____

Write the reciprocal of each of the following.

	a	*b*	*c*	*d*	*e*	*f*
3.	$\frac{5}{6}$ ___	$\frac{7}{8}$ ___	$\frac{1}{3}$ ___	6 ___	4 ___	8 ___
4.	$2\frac{1}{2}$ ___	$3\frac{1}{3}$ ___	$1\frac{1}{4}$ ___	$\frac{4}{9}$ ___	5 ___	$3\frac{1}{8}$ ___
5.	$4\frac{2}{3}$ ___	$\frac{1}{10}$ ___	12 ___	$1\frac{3}{5}$ ___	$\frac{7}{12}$ ___	$2\frac{13}{16}$ ___

Perfect score: 24   My score: _____

35

Lesson 9   Division

To divide by a fraction, multiply by its reciprocal.

Multiply by
the reciprocal.

$$\frac{3}{5} \div \frac{1}{2} = \frac{3}{5} \times \frac{2}{1}$$

$$= \frac{6}{5}$$

$$= 1\frac{1}{5}$$

Multiply by
the reciprocal.

$$1\frac{3}{5} \div 6 = \frac{8}{5} \times \frac{1}{6}$$

$$= \frac{4}{15}$$

Write each answer in simplest form.

	a	b	c	d
1.	$\frac{3}{4} \div \frac{1}{2}$	$\frac{1}{2} \div \frac{3}{4}$	$\frac{4}{5} \div \frac{8}{15}$	$\frac{7}{9} \div \frac{2}{3}$
2.	$\frac{7}{8} \div \frac{1}{2}$	$\frac{3}{4} \div \frac{1}{8}$	$\frac{4}{5} \div \frac{1}{2}$	$\frac{2}{3} \div \frac{1}{2}$
3.	$8 \div \frac{4}{5}$	$2\frac{1}{3} \div 2$	$2\frac{2}{5} \div 8$	$8\frac{1}{3} \div 10$
4.	$5\frac{1}{4} \div \frac{9}{20}$	$5\frac{1}{2} \div \frac{1}{4}$	$7\frac{3}{4} \div \frac{3}{16}$	$9 \div 6$
5.	$10 \div \frac{2}{5}$	$1\frac{7}{12} \div \frac{7}{12}$	$8 \div 12$	$3\frac{9}{10} \div \frac{13}{15}$

Perfect score: 20   My score: _____

36

Lesson 10   Division

$$3\frac{3}{8} \div 1\frac{1}{2} = \frac{27}{8} \div \frac{3}{2}$$ Change $3\frac{3}{8}$ and $1\frac{1}{2}$ to fractions.

$$= \frac{\overset{9}{\cancel{27}}}{\underset{4}{\cancel{8}}} \times \frac{\overset{1}{\cancel{2}}}{\underset{1}{\cancel{3}}}$$ Multiply by the reciprocal of $\frac{3}{2}$.

$$= \frac{9}{4} \text{ or } 2\frac{1}{4}$$ Write $\frac{9}{4}$ as a mixed numeral in simplest form.

Write each answer in simplest form.

	a	*b*	*c*	*d*
1.	$\frac{7}{10} \div \frac{14}{15}$	$2\frac{2}{3} \div \frac{2}{3}$	$\frac{2}{3} \div 2\frac{2}{3}$	$5 \div 1\frac{1}{4}$
2.	$2\frac{1}{3} \div 2\frac{1}{3}$	$2\frac{2}{3} \div 3\frac{1}{3}$	$1\frac{2}{5} \div 1\frac{2}{7}$	$2\frac{2}{5} \div 3\frac{1}{3}$
3.	$\frac{3}{5} \div 1\frac{2}{7}$	$1\frac{1}{8} \div \frac{5}{6}$	$1\frac{3}{7} \div \frac{5}{7}$	$\frac{1}{2} \div 1\frac{1}{4}$
4.	$3\frac{1}{3} \div 1\frac{1}{4}$	$1\frac{7}{8} \div 1\frac{1}{4}$	$1\frac{5}{12} \div 2\frac{2}{3}$	$5\frac{3}{5} \div 2\frac{1}{10}$

Perfect score: 16   My score: _____

Problem Solving

Solve. Write each answer in simplest form.

1. A leaky faucet wastes $\frac{1}{20}$ gallon of water each hour. In how many hours will $\frac{3}{4}$ gallon of water be used?

It will take _____ hours.

2. How many books can be placed on a shelf 14 inches long if each book is $\frac{7}{8}$ inch thick?

_____ books can be placed on the shelf.

3. Mr. Cepuran worked $6\frac{3}{4}$ hours giving piano lessons today. Each lesson was $\frac{3}{4}$ hour long. How many lessons did he give today?

He gave _____ lessons.

4. There are $5\frac{1}{4}$ quarts of water to be put into 3 cans. The same amount is to be put into each can. How many quarts of water will be in each can?

_____ quarts of water will be in each can.

5. Gina has a strip of felt $7\frac{7}{8}$ inches long. The felt can be cut into how many $1\frac{5}{16}$ inch pieces?

It can be cut into _____ pieces.

6. $1\frac{7}{10}$ gallons of gasoline were used to drive $25\frac{1}{2}$ miles. How many miles per gallon did the car get?

The car got _____ miles per gallon.

7. 20 pounds of soil are to be placed in containers. Each container will hold $3\frac{3}{4}$ pounds of soil. How many full containers will there be? How much of another container will be filled?

_____ containers will be filled.

_____ of another container will be filled.

1.
2.
3.
4.
5.
6.
7.

Perfect score: 8   My score: _____

38

Lesson 11   Addition and Subtraction

Study how to add or subtract when the denominators are the same.

Add the numerators.

$$\frac{2}{9} + \frac{5}{9} + \frac{8}{9} = \frac{2 + 5 + 8}{9}$$

Use the same denominator.

$$= \frac{15}{9}$$

$$= 1\frac{2}{3}$$

$\frac{2}{9}$
$\frac{5}{9}$
$+\frac{8}{9}$
$\frac{15}{9}$ or $1\frac{2}{3}$

Subtract the numerators.

$$\frac{7}{8} - \frac{3}{8} = \frac{7 - 3}{8}$$

Use the same denominator.

$$= \frac{4}{8}$$

$$= \frac{1}{2}$$

$\frac{7}{8}$
$-\frac{3}{8}$
$\frac{4}{8}$ or $\frac{1}{2}$

Write each answer in simplest form.

	a	b	c	d	e	f
1.	$\frac{3}{5}$ $+\frac{1}{5}$	$\frac{7}{8}$ $+\frac{6}{8}$	$\frac{3}{4}$ $+\frac{3}{4}$	$\frac{4}{5}$ $-\frac{3}{5}$	$\frac{5}{8}$ $-\frac{2}{8}$	$\frac{3}{8}$ $-\frac{1}{8}$
2.	$\frac{9}{10}$ $-\frac{1}{10}$	$\frac{3}{4}$ $-\frac{1}{4}$	$\frac{7}{8}$ $-\frac{1}{8}$	$\frac{7}{12}$ $-\frac{5}{12}$	$\frac{5}{6}$ $-\frac{3}{6}$	$\frac{7}{10}$ $-\frac{7}{10}$
3.	$\frac{1}{8}$ $\frac{3}{8}$ $+\frac{3}{8}$	$\frac{2}{7}$ $\frac{3}{7}$ $+\frac{4}{7}$	$\frac{7}{10}$ $\frac{9}{10}$ $+\frac{9}{10}$	$\frac{4}{15}$ $\frac{4}{15}$ $+\frac{7}{15}$	$\frac{7}{10}$ $\frac{2}{10}$ $+\frac{9}{10}$	$\frac{5}{12}$ $\frac{11}{12}$ $+\frac{11}{12}$

Perfect score: 18   My score: _____

Problem Solving

Solve. Write each answer in simplest form.

1. Mary Ann has a rock that weighs $\frac{3}{10}$ pound and Trudy has a rock that weighs $\frac{9}{10}$ pound. What is the combined weight of both rocks?

The combined weight is _____ pounds.

2. In problem **1**, how much more does the heavier rock weigh?

The heavier rock weighs _____ pound more.

3. Book A is $\frac{1}{8}$ inch thick, book B is $\frac{3}{8}$ inch thick, and book C is $\frac{5}{8}$ inch thick. What is the total thickness of all three books?

The total thickness is _____ inches.

4. In problem **3**, what is the difference in thickness between books A and C?

The difference is _____ inch.

5. In problem **3**, what is the difference in thickness between books B and C?

The difference is _____ inch.

6. Dwayne spent $\frac{3}{12}$ hour eating, $\frac{11}{12}$ hour watching TV, and $\frac{5}{12}$ hour reading. How much longer did Dwayne spend reading than eating?

He spent _____ hour more reading than eating.

7. In problem **6**, how much longer did he spend watching TV than reading?

He spent _____ hour more watching TV than reading.

8. In problem **6**, how much time did he spend in all three activities?

He spent _____ hours.

1.	
2.	
3.	
4.	
5.	
6.	
7.	
8.	

Perfect score: 8   My score: _____

40

Lesson 12 Renaming Numbers

You can rename a fraction by multiplying the numerator and the denominator by the same number.

$$\frac{3}{4} = \frac{}{20}$$

$$\frac{3}{4} = \frac{3 \times 5}{4 \times 5}$$

$$= \frac{15}{20}$$

Choose 5 so the new denominator is 20.

$$4\frac{2}{5} = 4\frac{}{15}$$

$$4\frac{2}{5} = 4\frac{2 \times 3}{5 \times 3}$$

$$= 4\frac{6}{15}$$

Choose 3 so the new denominator is 15.

Rename.

	a	*b*	*c*
1.	$\frac{1}{2} = \frac{}{16}$	$\frac{3}{4} = \frac{}{12}$	$\frac{3}{5} = \frac{}{15}$
2.	$\frac{4}{5} = \frac{}{10}$	$\frac{7}{9} = \frac{}{45}$	$\frac{5}{12} = \frac{}{36}$
3.	$\frac{7}{8} = \frac{}{24}$	$\frac{1}{6} = \frac{}{30}$	$\frac{5}{12} = \frac{}{60}$
4.	$1\frac{2}{3} = 1\frac{}{6}$	$2\frac{5}{8} = 2\frac{}{40}$	$4\frac{1}{4} = 4\frac{}{8}$
5.	$3\frac{8}{9} = 3\frac{}{18}$	$6\frac{7}{10} = 6\frac{}{60}$	$7\frac{5}{6} = 7\frac{}{24}$

Perfect score: 15   My score: _____

Lesson 13   Addition and Subtraction

To add or subtract when the denominators are different, rename the fractions so the denominators are the same.

The denominators are 4 and 5. Since $4 \times 5 = 20$, rename each fraction with a denominator of 20.

$$\frac{3}{4} \rightarrow \frac{15}{20}$$
$$+\frac{3}{5} \rightarrow +\frac{12}{20}$$
$$\frac{27}{20} = 1\frac{7}{20}$$

$$\frac{2}{5} \rightarrow \frac{8}{20}$$
$$-\frac{1}{4} \rightarrow -\frac{5}{20}$$
$$\frac{3}{20}$$

The denominators are 2 and 3. Since $2 \times 3 = 6$, rename each number with a denominator of 6.

$$3\frac{2}{3} \rightarrow 3\frac{4}{6}$$
$$+1\frac{1}{2} \rightarrow +1\frac{3}{6}$$
$$4\frac{7}{6} = 5\frac{1}{6}$$

$$5\frac{1}{2} \rightarrow 5\frac{3}{6}$$
$$-2\frac{1}{3} \rightarrow -2\frac{2}{6}$$
$$3\frac{1}{6}$$

— Write answers in simplest form. —

Write each answer in simplest form.

	a	b	c	d
1.	$\frac{3}{5}$ $+\frac{2}{3}$	$\frac{3}{4}$ $+\frac{1}{3}$	$\frac{7}{8}$ $+\frac{1}{3}$	$\frac{1}{2}$ $+\frac{2}{3}$
2.	$\frac{1}{2}$ $-\frac{2}{5}$	$\frac{2}{3}$ $-\frac{1}{5}$	$\frac{7}{8}$ $-\frac{2}{3}$	$\frac{3}{4}$ $-\frac{2}{3}$
3.	$1\frac{3}{4}$ $+3\frac{2}{5}$	$3\frac{5}{8}$ $+2\frac{4}{5}$	$4\frac{1}{2}$ $+1\frac{2}{3}$	$5\frac{2}{3}$ $+2\frac{3}{10}$
4.	$3\frac{2}{3}$ $-1\frac{1}{2}$	$5\frac{1}{5}$ $-2\frac{1}{6}$	$4\frac{3}{4}$ $-4\frac{2}{3}$	$7\frac{4}{5}$ $-6\frac{7}{12}$

Perfect score: 16   My score: _____

Lesson 14 Addition and Subtraction

**When one denominator is a factor of the other, use
the greater denominator as the common denominator.**

2 is a factor of 10, so use
10 as the common denominator.

$$\frac{7}{10} \rightarrow \frac{7}{10}$$
$$+\frac{1}{2} \rightarrow +\frac{5}{10}$$
$$\frac{12}{10} = 1\frac{1}{5}$$

$$3\frac{1}{2} \rightarrow 3\frac{5}{10}$$
$$+1\frac{3}{10} \rightarrow +1\frac{3}{10}$$
$$4\frac{8}{10} = 4\frac{4}{5}$$

3 is a factor of 6, so use
6 as the common denominator.

$$\frac{2}{3} \rightarrow \frac{4}{6}$$
$$-\frac{1}{6} \rightarrow -\frac{1}{6}$$
$$\frac{3}{6} = \frac{1}{2}$$

$$4\frac{5}{6} \rightarrow 4\frac{5}{6}$$
$$-1\frac{2}{3} \rightarrow -1\frac{4}{6}$$
$$3\frac{1}{6}$$

Write each answer in simplest form.

	a	b	c	d

1.
a. $\frac{1}{2}$ $+\frac{3}{4}$
b. $\frac{7}{8}$ $+\frac{3}{4}$
c. $\frac{8}{9}$ $+\frac{1}{3}$
d. $\frac{2}{3}$ $+\frac{1}{12}$

2.
a. $3\frac{2}{5}$ $+1\frac{7}{20}$
b. $6\frac{11}{16}$ $+3\frac{1}{2}$
c. $7\frac{7}{24}$ $+4\frac{5}{6}$
d. $5\frac{1}{4}$ $+2\frac{5}{12}$

3.
a. $\frac{9}{10}$ $-\frac{1}{2}$
b. $\frac{5}{6}$ $-\frac{1}{3}$
c. $\frac{7}{8}$ $-\frac{1}{2}$
d. $\frac{3}{4}$ $-\frac{7}{16}$

4.
a. $2\frac{5}{6}$ $-1\frac{2}{3}$
b. $6\frac{5}{8}$ $-3\frac{1}{4}$
c. $5\frac{4}{5}$ $-2\frac{9}{20}$
d. $9\frac{11}{24}$ $-7\frac{3}{8}$

Perfect score: 16   My score: _____

Problem Solving

Stock-Market Report for McTavish Dog Supplies
(changes given in dollars)

Monday	Tuesday	Wednesday	Thursday	Friday
up $\frac{7}{8}$	up $\frac{3}{4}$	up $1\frac{1}{2}$	up $2\frac{7}{8}$	

Solve. Write each answer in simplest form.

1. How much greater was Monday's gain than Tuesday's?

Monday's gain was _____ dollar greater.

2. What was the combined gain for Monday and Tuesday?

The combined gain was _____ dollars.

3. What was the combined gain for Wednesday and Thursday?

The combined gain was _____ dollars.

4. The price before trading Monday was $23\frac{5}{8}$ dollars. What was the price after Monday's trading?

The price was _____ dollars.

5. The price after Thursday's trading was $29\frac{5}{8}$ dollars. The price after Friday's trading was $26\frac{1}{2}$ dollars. How much did the stock go down on Friday?

It went down _____ dollars.

6. A recipe calls for $\frac{3}{4}$ cup flour and $\frac{2}{3}$ cup ground nuts. How much more flour does the recipe call for than ground nuts?

The recipe calls for _____ cup more flour.

7. A gallon of water is poured into a jug that weighs $2\frac{9}{10}$ pounds. The water weighs $8\frac{1}{3}$ pounds. What is the combined weight of the water and the jug?

The combined weight is _____ pounds.

1.

2.

3.

4.

5.

6.

7.

Perfect score: 7   My score: _____

44

Lesson 15   Addition and Subtraction

The denominators 4 and 10 have a common factor of 2. Use $(4 \times 10) \div 2$ or 20 as the common denominator.

$$\frac{3}{4} \to \frac{15}{20}$$
$$+\frac{7}{10} \to +\frac{14}{20}$$
$$\frac{29}{20} = 1\frac{9}{20}$$

$$2\frac{9}{10} \to 2\frac{18}{20}$$
$$+3\frac{1}{4} \to +3\frac{5}{20}$$
$$5\frac{23}{20} = 6\frac{3}{20}$$

The denominators 8 and 12 have a common factor of 4. Use $(8 \times 12) \div 4$ or 24 as the common denominator.

$$\frac{7}{8} \to \frac{21}{24}$$
$$-\frac{5}{12} \to -\frac{10}{24}$$
$$\frac{11}{24}$$

$$6\frac{11}{12} \to 6\frac{22}{24}$$
$$-2\frac{5}{8} \to -2\frac{15}{24}$$
$$4\frac{7}{24}$$

Write each answer in simplest form.

	a	b	c	d
1.	$\frac{1}{6}$ $+\frac{3}{8}$	$\frac{3}{4}$ $+\frac{1}{6}$	$\frac{3}{10}$ $+\frac{4}{15}$	$\frac{5}{6}$ $+\frac{4}{9}$
2.	$1\frac{3}{4}$ $+1\frac{3}{10}$	$3\frac{7}{15}$ $+2\frac{1}{6}$	$4\frac{11}{12}$ $+5\frac{8}{9}$	$8\frac{5}{6}$ $+4\frac{7}{10}$
3.	$\frac{1}{4}$ $-\frac{1}{6}$	$\frac{8}{9}$ $-\frac{5}{6}$	$\frac{7}{12}$ $-\frac{3}{16}$	$\frac{13}{25}$ $-\frac{7}{15}$
4.	$2\frac{5}{6}$ $-1\frac{1}{4}$	$6\frac{11}{15}$ $-3\frac{7}{10}$	$9\frac{1}{8}$ $-5\frac{1}{10}$	$8\frac{17}{20}$ $-2\frac{5}{12}$

Perfect score: 16   My score: _____

Problem Solving

Solve. Express each answer in simplest form.

1. Percy worked $\frac{5}{6}$ hour on Monday, $\frac{3}{4}$ hour on Tuesday, and $\frac{9}{10}$ hour on Wednesday. How many hours did he work on Monday and Tuesday?

He worked _____ hours on Monday and Tuesday.

2. In problem **1**, did Percy work longer on Monday or Wednesday? How much longer did he work on that day?

He worked longer on _____.

He worked _____ hour longer.

3. In problem **1**, how many hours did Percy work on Tuesday and Wednesday?

He worked _____ hours on all three days.

4. The blue tape is $\frac{9}{16}$ inch wide, the red tape is $1\frac{7}{8}$ inches wide, and the green tape is $1\frac{3}{4}$ inches wide. What is the combined width of the blue and the green tapes?

The combined width is _____ inches.

5. In problem **4**, is the red tape or green tape wider? How much wider is it?

The _____ tape is wider.

It is _____ inch wider.

6. On a test, Brenda worked $\frac{7}{30}$ hour, Emma worked $\frac{7}{20}$ hour, and Laura worked $\frac{7}{15}$ hour. Which girl worked the least amount of time? How long did the other two girls work on the test?

_____ worked the least amount of time.

The other girls worked _____ hour.

7. In problem **6**, how long did Brenda and Emma work on the test?

They worked _____ hours.

1.	
2.	**3.**
4.	**5.**
6.	**7.**

Perfect score: 10   My score: _____

Lesson 16 Subtraction

To subtract, rename 6 as $5\frac{4}{4}$.

$$6 \quad\rightarrow\quad 5\frac{4}{4}$$
$$-1\frac{3}{4} \qquad -1\frac{3}{4}$$
$$\overline{} \qquad \overline{4\frac{1}{4}}$$

$$\begin{aligned} 6 &= 5 + 1 \\ &= 5 + \frac{4}{4} \\ &= 5\frac{4}{4} \end{aligned}$$

To subtract, rename $4\frac{5}{15}$ as $3\frac{20}{15}$.

$$4\frac{1}{3} \quad\rightarrow\quad 4\frac{5}{15} \quad\rightarrow\quad 3\frac{20}{15}$$
$$-1\frac{3}{5} \qquad -1\frac{9}{15} \qquad -1\frac{9}{15}$$
$$\overline{} \qquad \overline{} \qquad \overline{2\frac{11}{15}}$$

$$\begin{aligned} 4\frac{5}{15} &= 3 + 1 + \frac{5}{15} \\ &= 3 + \frac{15}{15} + \frac{5}{15} \\ &= 3\frac{20}{15} \end{aligned}$$

Write each answer in simplest form.

	a	b	c	d

1.
a) 7 $-\frac{7}{8}$

b) 9 $-\frac{3}{10}$

c) 6 $-3\frac{1}{2}$

d) $4\frac{2}{7}$ $-2\frac{6}{7}$

2.
a) $4\frac{1}{4}$ $-2\frac{1}{2}$

b) $4\frac{2}{3}$ $-1\frac{4}{5}$

c) $6\frac{1}{2}$ $-1\frac{3}{8}$

d) $1\frac{1}{3}$ $-\frac{1}{2}$

3.
a) 2 $-\frac{3}{4}$

b) $3\frac{1}{2}$ $-1\frac{2}{3}$

c) 6 $-1\frac{5}{6}$

d) $7\frac{1}{8}$ $-4\frac{5}{6}$

4.
a) $5\frac{2}{3}$ $-\frac{5}{6}$

b) $4\frac{2}{15}$ $-3\frac{7}{10}$

c) $7\frac{5}{9}$ $-4\frac{7}{12}$

d) $8\frac{3}{4}$ $-3\frac{4}{5}$

Perfect score: 16   My score: _____

Problem Solving

Rehearsal Schedule	
Day	*Time*
Monday	2 hours
Tuesday	$1\frac{5}{6}$ hours
Wednesday	$\frac{3}{4}$ hour
Thursday	$1\frac{3}{5}$ hours
Friday	

Solve each problem.

1. How many hours less did the group practice on Wednesday than on Monday?

They practiced _____ hours less on Wednesday.

2. How much longer did they practice on Tuesday than on Thursday?

They practiced _____ hour longer on Tuesday.

3. Find the combined practice time for Wednesday and Thursday.

The combined practice time was _____ hours.

4. Find the average time they practiced on Wednesday and Thursday.

The average practice time was _____ hours.

5. On Friday they practiced twice as long as they practiced on Wednesday. How long did they practice on Friday?

They practiced _____ hours on Friday.

1.

2.

3.

4.

5.

Perfect score: 5   My score: _____

CHAPTER 2 TEST

Write each answer in simplest form.

a	b	c	d

1. $\dfrac{2}{3} \times \dfrac{2}{3}$ $\qquad$ $\dfrac{4}{5} \times \dfrac{5}{6}$ $\qquad$ $\dfrac{3}{10} \times \dfrac{8}{9}$ $\qquad$ $\dfrac{1}{2} \times \dfrac{2}{3} \times \dfrac{3}{4}$

2. $12 \times 7\dfrac{1}{2}$ $\qquad$ $1\dfrac{4}{5} \times 3\dfrac{1}{3}$ $\qquad$ $\dfrac{8}{9} \times 7\dfrac{1}{2}$ $\qquad$ $4 \times 3\dfrac{1}{2} \times 2\dfrac{3}{4}$

3. $\dfrac{4}{5} \div \dfrac{5}{6}$ $\qquad$ $\dfrac{7}{8} \div \dfrac{3}{10}$ $\qquad$ $12 \div 1\dfrac{1}{8}$ $\qquad$ $4\dfrac{2}{3} \div 1\dfrac{1}{6}$

4. $\begin{array}{r} \dfrac{2}{5} \\ +\dfrac{2}{5} \\ \hline \end{array}$ $\qquad$ $\begin{array}{r} \dfrac{7}{9} \\ +\dfrac{2}{3} \\ \hline \end{array}$ $\qquad$ $\begin{array}{r} 3\dfrac{3}{4} \\ +4\dfrac{1}{3} \\ \hline \end{array}$ $\qquad$ $\begin{array}{r} \dfrac{5}{6} \\ +2\dfrac{3}{10} \\ \hline \end{array}$

5. $\begin{array}{r} \dfrac{9}{10} \\ -\dfrac{3}{10} \\ \hline \end{array}$ $\qquad$ $\begin{array}{r} \dfrac{6}{7} \\ -\dfrac{3}{8} \\ \hline \end{array}$ $\qquad$ $\begin{array}{r} 7 \\ -2\dfrac{1}{4} \\ \hline \end{array}$ $\qquad$ $\begin{array}{r} 3\dfrac{2}{5} \\ -1\dfrac{7}{10} \\ \hline \end{array}$

Perfect score: 20   My score: _____

PRE-TEST—Operations Involving Decimals

Add or subtract.

	a	b	c	d	e
1.	4.2 +5.3	1.2 7 8 +.8 3 1	4 5.8 9 +2.6	4.5 6 +.3 8 9	2 4 +.8 3 1
2.	.7 3 −.2 4	4 1.3 8 −2.4 7	6.4 1 3 −1.2 8	4 2.1 −1.6 8 7	3 8 −6.2 4
3.	1.4 2 4 .6 8 1 +.7 3 2	3.8 2 1 4.6 8 3 +.8	1 8 3.6 +7 0.2 1 9	5 2.7 4 −9.3 8	6 8.2 −1.5 3 9

Multiply or divide.

	a	b	c	d
4.	5.8 ×.6	7 3 ×.0 2	.4 9 ×.0 0 8	1 3.2 ×.0 7
5.	4⟌7.6	6⟌.0 2 2 8	.6⟌1 8 9	.08⟌.4 3 2
6.	.6 8 3 ×3 2.4	7 2.0 8 ×.0 4 2	4.2⟌1.5 5 4	.26⟌9 6.2

Perfect score: 27   My score: _____

Lesson 1 Decimals, Fractions, and Mixed Numerals

> Any fraction that can be renamed as *tenths*, *hundredths*, *thousandths*, and so on, can be changed to a decimal.

$\frac{1}{10} = .1$ $\qquad$ $\frac{3}{10} = .3$ $\qquad$ $5\frac{7}{10} = 5.7 \rightarrow$ *five and seven tenths*

$\frac{1}{100} = .01$ $\qquad$ $\frac{17}{100} = .17$ $\qquad$ $8\frac{9}{100} = 8.09 \rightarrow$ *eight and nine hundredths*

$\frac{1}{1000} = .001$ $\qquad$ $\frac{153}{1000} = .153$ $\qquad$ $17\frac{11}{1000} = 17.011 \rightarrow$ *seventeen and eleven thousandths*

Write each of the following as a decimal.

	a	*b*	*c*
1.	$\frac{9}{10} =$ _____	$3\frac{9}{10} =$ _____	$\frac{7}{10} =$ _____
2.	$\frac{17}{100} =$ _____	$3\frac{53}{100} =$ _____	$\frac{7}{100} =$ _____
3.	$\frac{259}{1000} =$ _____	$4\frac{357}{1000} =$ _____	$\frac{7}{1000} =$ _____
4.	$5\frac{5}{10} =$ _____	$12\frac{40}{100} =$ _____	$4\frac{14}{100} =$ _____
5.	$11\frac{1}{10} =$ _____	$39\frac{28}{1000} =$ _____	$25\frac{3}{100} =$ _____

Write each decimal as a fraction or as a mixed numeral.

6.	.9 = _____	.3 = _____	4.1 = _____
7.	.19 = _____	.07 = _____	5.03 = _____
8.	.419 = _____	.011 = _____	3.333 = _____
9.	13.3 = _____	20.27 = _____	20.027 = _____
10.	9.03 = _____	100.1 = _____	4.567 = _____

Perfect score: 30 $\qquad$ My score: _____

Decimals, Fractions, and Mixed Numerals

Change $\frac{3}{5}$ to *tenths*.

$$\frac{3}{5} = \frac{3 \times 2}{5 \times 2}$$
$$= \frac{6}{10}$$
$$= .6$$

Change $3\frac{1}{4}$ to *hundredths*.

$$3\frac{1}{4} = 3 + \frac{1 \times 25}{4 \times 25}$$
$$= 3 + \frac{25}{100}$$
$$= 3\frac{25}{100}$$
$$= 3.25$$

Change $\frac{9}{20}$ to *thousandths*.

$$\frac{9}{20} = \frac{9 \times 50}{20 \times 50}$$
$$= \frac{450}{1000}$$
$$= .450$$

Any decimal can be changed to a fraction or a mixed numeral.

$$.8 = \frac{8}{10}$$
$$= \frac{4}{5}$$

$$4.75 = 4\frac{75}{100}$$
$$= 4\frac{3}{4}$$

$$.925 = \frac{925}{1000}$$
$$= \frac{37}{40}$$

Change each of the following to a decimal as indicated.

a	*b*	*c*
Change to *tenths*.	Change to *hundredths*.	Change to *thousandths*.
1. $\frac{1}{5} =$ _____	$\frac{3}{4} =$ _____	$\frac{9}{500} =$ _____
2. $3\frac{4}{5} =$ _____	$5\frac{7}{10} =$ _____	$1\frac{17}{250} =$ _____
3. $5\frac{1}{2} =$ _____	$4\frac{12}{25} =$ _____	$9\frac{123}{200} =$ _____
4. $12\frac{1}{2} =$ _____	$3\frac{9}{20} =$ _____	$4\frac{7}{100} =$ _____

Change each decimal to a fraction or a mixed numeral in simplest form.

5. $.7 =$ _____	$.4 =$ _____	$6.5 =$ _____
6. $.21 =$ _____	$.02 =$ _____	$1.75 =$ _____
7. $.213 =$ _____	$.050 =$ _____	$5.555 =$ _____
8. $.003 =$ _____	$6.09 =$ _____	$10.108 =$ _____

Perfect score: 24   My score: _____

Lesson 2   Addition

NAME _____

Keep the decimal points aligned.
Add as you would add whole numbers.

```
  1 1 1                    1 1                      1 1
1 4.8 1 4                1.2 4                     1.2 4 0
  3.5 5 1                4 5.3         or          4 5.3 0 0    ←  Write these 0's
+  .4 3 6              + 2.5 8 7                  + 2.5 8 7        if they help you.
─────────             ─────────                  ─────────
1 8.8 0 1             4 9.1 2 7                   4 9.1 2 7
```

Put a decimal point in the answer.

Add.

	a	*b*	*c*	*d*	*e*
1.	.3 +.5	.6 +.5	4.6 +2.9	5.8 +7.6	3.4 +9.8
2.	.2 4 +.6 2	.0 3 +.4 0	$ .5 8 +.2 7	$ 4.5 6 +2.8 3	$ 1 4.3 8 +2 9.7 5
3.	.0 4 2 +.1 5 3	.2 4 6 +.1 7 3	.4 2 6 +.1 7 9	3.8 2 4 +6.2 9 2	4 2.3 7 5 +8 4.9 6 8
4.	.9 +.8 5	.1 2 7 +.5 6	3.9 6 5 +4.0 7	1 7.3 6 4 +4.9	5.3 0 8 +9 2.6 7
5.	.3 0 8 +.6	.4 6 +.2 0 7	6.0 5 +1 8.9	3.2 +.0 6 9	4 2.0 5 +2.7 8 3
6.	.1 2 6 .2 +.8 4	.8 9 .3 +.9 2 6	.4 6 3 4.2 +9.3 8 4	.9 1 4.6 2 5 +3.0 2	.4 2 4 2.3 +9.6 1 8

Perfect score: 30   My score: _____

53

Problem Solving

Solve each problem.

1. One package weighs .8 kilogram. Another weighs .6 kilogram. What is the combined weight of these two packages?

The combined weight is _____ kilograms.

2. The thickness of a board is .037 meter. This is .014 meter less than what it is supposed to be. How thick is it supposed to be?

It is supposed to be _____ meter thick.

3. The odometer readings on three cars are 4,216.7; 382.4; and 53,318.6. According to the odometers, what is the total number of miles the cars were driven?

The cars were driven _____ miles.

4. David made purchases at three stores. The amounts were $1.58; $3.97; and $.97. What was the total amount of his purchases?

The total amount was $_____.

5. Consider the numbers named by .328; 32.8; 3.28; and 328. What is the sum of the two greatest numbers?

The sum is _____.

6. In problem 5, what is the sum of the two least numbers?

The sum is _____.

7. In problem 5, what is the sum of all four of the numbers?

The sum is _____.

8. Able and Mable were asked to find the sum of 4.26; .364; 37; and .5. Able said the sum is 41.629 and Mable said the sum is 42.124. Who is correct?

_____ is correct.

1.	2.
3.	**4.**
5.	**6.**
7.	**8.**

Perfect score: 8   My score: _____

Lesson 3   Subtraction

Keep the decimal points aligned.
Subtract as you would subtract whole numbers.

```
    3 13
3 7.4 3 8          2 5.              2 5.0 0 0  ←— Write these 0's
-2 5.3 6 5        - 1.0 7 5          - 1.0 7 5     if they help you.
─────────   or   ─────────          ─────────
1 2.0 7 3          2 3.9 2 5          2 3.9 2 5
```

Put a decimal point in the answer.

Subtract.

	a	*b*	*c*	*d*	*e*
1.	.9 −.3	.4 7 −.1 8	.9 2 4 −.8 3 1	4.7 3 4 −2.6 8 5	5 2.6 1 3 −1 4.0 8 2
2.	.7 3 6 −.1 4 2	4.3 5 −2.9 6	$ 1 3.0 8 −1 2.5 0	3.4 1 6 −2.0 3 7	$ 4 2 3.2 5 −1 2 6.8 3
3.	.3 5 −.2	.5 1 6 −.3 8	4.2 6 −.4	1 2.5 6 3 −4.2 7	5 6 3.0 2 −3 4.6
4.	.1 2 6 −.0 4	1.3 7 −.4	3 2.6 5 −1.7	3 4.2 9 −2 1.6	4 2.3 1 8 −1 0.2 4
5.	8 −3.6	9.5 −2.6 1	9.8 −3.2 1 7	4 2.8 3 −1.0 4 6	6 5 −1.2 7 5
6.	.6 −.3 8 4	7 −.2 3 8	4.6 −1.2 7 3	3 4.2 −.0 0 8	3 6.2 −.6 2 5
7.	.4 −.3 1 8	2.2 6 −.4 2 3	6.0 8 2 −.1 4	7.1 1 2 −.4	4 2.0 8 −4 0.0 0 6

Perfect score: 35   My score: _____

Problem Solving

Solve each problem.

1. Naomi has two packages that weigh a total of 4.8 kilograms. One package weighs 1.9 kilograms. How much does the other package weigh?

The other package weighs _____ kilograms.

2. Fred needs $1.43 to purchase a model car. He now has $.75. How much more money does he need in order to buy the model car?

He needs $_____ more.

3. The average yearly rainfall for Will County is 35.50 inches. Last year 29.75 inches of rain were recorded. How many inches below the average was this?

It was _____ inches below the average.

4. Two sheets of plastic have a combined thickness of 1.080 centimeters. One sheet is .675 centimeter thick. What is the thickness of the other sheet?

The other sheet is _____ centimeter thick.

5. Jane gave the clerk a $20 bill. The amount of her purchases was $17.14. How much change should she receive?

She should receive $_____ change.

6. George Foster had a batting average of .323 one season. Pete Rose's average was .306 that season. How much better was George Foster's average?

Foster's average was _____ better.

7. How many gallons were purchased in all?

_____ gallons were purchased.

8. How many more gallons were purchased at the 1st stop than at the 3d stop?

_____ more gallons were purchased.

Gasoline Purchased During a Trip	
stops	*gallons*
1st	12
2d	8.7
3d	8.4

1.	2.
3.	4.
5.	6.
7.	8.

Perfect score: 8   My score: _____

56

Lesson 4   Addition and Subtraction

To find the sum, rewrite
213 + 4.2 + .635 + 13.54 as
follows. Then add.

$$
\begin{array}{r}
2\overset{1}{1}\overset{1}{3} \\
4.2 \\
.6\,3\,5 \\
+\,1\,3.5\,4 \\
\hline
2\,3\,1.3\,7\,5
\end{array}
$$

213 + 4.2 + .635 + 13.54 = _____

To find the difference,
rewrite 4.3 − .168 as follows.
Then subtract.

$$
\begin{array}{r}
4.3\overset{2}{\cancel{0}}\overset{9}{\cancel{0}}\overset{10}{\cancel{0}} \\
-.1\,6\,8 \\
\hline
4.1\,3\,2
\end{array}
$$

4.3 − .168 = _____

Add or subtract.

1.   14.3 + 2.687 = _____

2.   5.08 − 1.2 = _____

3.   15.92 + .038 = _____

4.   4.608 − 2.9 = _____

5.   14.213 + 1.425 + 3.16 = _____

6.   17 − 1.835 = _____

7.   3.082 + 16.4 + 6.42 = _____

8.   4.3 − .987 = _____

9.   4.2 + 38 + .04 + 1.163 = _____

10.   5.62 − 1.8 = _____

11.   .54 + 2.8 + 3.017 + .24 = _____

12.   17.93 − 8 = _____

Perfect score: 12   My score: _____

Problem Solving

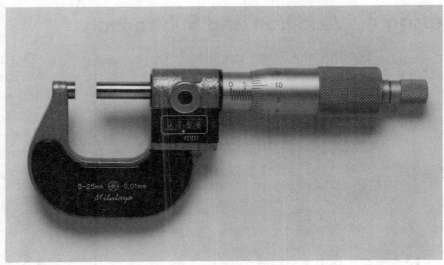

bar	measurement
A	.986 millimeter
B	.432 millimeter
C	1.035 millimeters
D	1.008 millimeters

Solve each problem.

1. The workers are using micrometers to measure the thickness of metal bars. They have recorded some measurements on the chart. What is the combined thickness of bars **A** and **B**?

The combined thickness is _____ millimeters.

2. Suppose the workers place the thickest and the thinnest bars next to each other. What would the combined thickness be?

It would be _____ millimeters.

3. How much thicker is bar **D** than bar **B**?

Bar **D** is _____ millimeter thicker.

4. The thickest bar is how much thicker than the thinnest bar?

The thickest bar is _____ millimeter thicker.

5. What is the total thickness of all four bars?

The total thickness is _____ millimeters.

6. Bar **A** is placed next to bar **C**, and bar **B** is placed next to bar **D**. Is the total thickness of **A** and **C** more or less than the total thickness of **B** and **D**? By how much?

The thickness of **A** and **C** is _____ than the thickness of **B** and **D**.

The difference is _____ millimeter.

1.	2.
3.	4.
5.	6.

Perfect score: 7   My score: _____

58

Lesson 5 Multiplication

number of digits after the decimal point

321	3.21	2	3.21	2	.321	3
×4	×4	+0	×.4	+1	×.04	+2
1284	12.84	2	1.284	3	.01284	5

$12 \times 27 = 324$, so $12 \times 2.7 =$ _____

and $12 \times .027 =$ _____.

$22 \times 18 = 396$, so $2.2 \times 1.8 =$ _____

and $.022 \times 1.8 =$ _____.

Multiply.

	a	b	c	d	e
1.	.2 3 ×6	.4 0 ×2	1.6 ×4	.3 2 4 ×5	4 3.1 ×7
2.	3 2 ×.3	4 2 ×.8	5.1 ×.2	.7 3 4 ×.9	2.1 6 ×.6
3.	.2 3 ×.0 4	.1 7 ×.0 5	2.5 ×.0 3	7.0 9 ×.0 2	.8 6 3 ×.0 4
4.	.4 2 ×.0 0 6	3.8 ×.0 0 8	.7 6 ×.0 0 4	5 1.2 ×.0 0 4	8.0 6 ×.0 0 3
5.	3 1 6 ×.0 3	2.0 4 ×.7	.7 1 5 ×.9	3 1.8 ×.0 4	6.1 9 ×.0 5
6.	4.2 1 ×.3	6 1.3 ×.0 8	3 0.8 ×.0 0 9	2.0 9 ×.0 2	4 1.6 ×.0 3
7.	3.2 4 ×5	6 8 7 ×.0 3	4.1 8 ×.0 9	3 1 7 ×.8	.6 1 7 ×4

Perfect score: 35   My score: _____

Multiplication

Multiply.

	a	*b*	*c*	*d*	*e*
1.	2.5 ×.3 7	4 2.1 ×.3 8	3 1 6.4 ×2.6	2.1 6 ×2 4.2	.0 4 2 1 ×3 2.1
2.	4 3 ×.2 8	4 8.6 ×3.1	.0 3 8 2 ×4 1	.3 1 8 ×4.1 2	.0 3 1 6 ×1 1.2
3.	.3 1 ×.1 6	7.3 1 ×.2 4	4 2.1 0 ×.1 3	3 0.8 ×.1 4 1	.0 4 1 2 ×3 0.6
4.	8.4 ×9.2	.5 1 3 ×2.6	3.1 6 0 ×4.3	2.0 4 ×1 0.5	1 2.1 1 ×4 0.1

5. In 1 hour, 18.5 tons of ore can be processed. At that rate, how much ore can be processed in 7.5 hours?

_____tons would be processed.

6. In problem 5, how many tons would be processed in .75 hour?

_____ tons would be processed.

7. What is the product of 42; .06; and 1.3?

The product is _____.

5.	
6.	**7.**

Perfect score: 23   My score: _____

60

Lesson 6 Multiplication

74.2	74.2	74.2
×10	×100	×1000
742.0	7420.0	74200.0
or	or	or
742	7420	74200

Shortcut

74.2 × 10 = 74 2
74.2 × 100 = 74 20
74.2 × 1000 = 74 200

.568	.568	.568
×10	×100	×1000
5.680	56.800	568.000
or	or	or
5.68	56.8	568

.568 × 10 = 5.68
.568 × 100 = 56.8
.568 × 1000 = 568

To multiply by 10, move the decimal point ___1___ place to the right.

To multiply by 100, move the decimal point _____ places to the right.

To multiply by 1000, move the decimal point _____ places to the right.

Complete the table.

	Number	*a* Multiply by 10.	*b* Multiply by 100.	*c* Multiply by 1000.
1.	8.3			
2.	.083			
3.	75.4			
4.	6.03			
5.	.0345			
6.	100.25			
7.	2.9064			

Perfect score: 21   My score: _____

61

Problem Solving

Solve each problem.

1. There is .001 gram of iron in one egg. How much iron is there in 3 eggs?

There is _____ gram of iron in 3 eggs.

2. A liter of oil weighs about .804 kilogram. How many kilograms would 9.5 liters of oil weigh?

It would weigh _____ kilograms.

3. A gallon of water weighs about 8.34 pounds. Melissa used 21.5 gallons of water when she took a bath. What was the weight of the water she used?

The water weighed _____ pounds.

4. Milo's car can be driven an average of 21.6 miles on each gallon of gasoline. How far can his car be driven on 9.8 gallons of gasoline?

His car can be driven _____ miles.

5. Each box of cereal weighs .45 kilogram. What is the weight of 10 boxes? What is the weight of 100 boxes?

The weight of 10 boxes is _____ kilograms.

The weight of 100 boxes is _____ kilograms.

6. A jet traveled at an average speed of 681 kilometers an hour. At that rate, how far did the jet go in 7.5 hours?

The jet traveled _____ kilometers.

7. Brigida earns $5.54 an hour. How much will she earn in 10 hours? How much will she earn in 100 hours?

She will earn $_____ in 10 hours.

She will earn $_____ in 100 hours.

1.	
2.	**3.**
4.	**5.**
6.	**7.**

Perfect score: 9   My score: _____

Lesson 7   Division

When dividing a number like 2.8; .28; or .028 by any whole number, place the decimal point in the quotient directly above the decimal point in the dividend.

Check		*Check*		*Check*

```
    .4                      .04                      .004
7)2.8       7          7).28        7          7).028       7
  28      × .4           28       × .04          28      × .004
 ───      ────          ───       ────          ───      ─────
   0      2.8             0        .28             0       .028
```

$65 \div 13 = 5$, so $6.5 \div 13 =$ _____; $.65 \div 13 =$ _____;   and $.065 \div 13 =$ _____.

Divide.

	a	*b*	*c*	*d*
1.	4).6 4	2)3.6	6)7 4.4	3).9 5 4
2.	5).4 5	7).0 5 6	4).0 9 2	3)6.0 8 1
3.	3).0 1 2	2).0 0 8	9).0 7 2	4)3 0.7 2
4.	3).0 5 4	4).0 6 8	3).0 8 4	6).0 0 8 4

Perfect score: 16   My score: _____

63

Problem Solving

Solve each problem.

1. A case of tacks weighs .294 kilogram. There are 7 boxes of tacks in a case, each box having the same weight. Find the weight of each box.

Each box weighs _____ kilogram.

2. It costs $4.32 to mail a package that weighs 3 kilograms. What is the average cost per kilogram?

The average cost per kilogram is $_____.

3. A wire 15.3 meters long is to be separated into 9 pieces. Each piece is to be the same length. How long will each piece of wire be?

Each piece will be _____ meters long.

4. Six sheets of paper have a total thickness of .072 inch. Each sheet has the same thickness. How thick is each sheet?

Each sheet is _____ inch thick.

5. In 5 hours, 1.85 tons of ore were processed. The same amount was processed each hour. How many tons were processed each hour?

_____ ton was processed each hour.

6. There are 3.76 liters of liquid to be put into 4 containers. The same amount is to be put into each one. How much liquid will be put into each container?

_____ liter will be put into each container.

7. The barometric pressure rose 1.44 inches in 3 hours. What was the average change per hour?

The average change per hour was _____ inch.

1.	
2.	**3.**
4.	**5.**
6.	**7.**

Lesson 8 Division

$$.9\overline{)\,.45}$$ →

| To get a whole number divisor, multiply both .9 and .45 by 10. | → | $$\begin{array}{r}.5\\9\overline{)4.5}\\45\\\hline 0\end{array}$$ | *shorter way*
$$\begin{array}{r}.5\\.9\overline{)\,.45}\\45\\\hline 0\end{array}$$ |

Check: $.5 \times .9 =$ _____

$$.03\overline{)4.8}$$ →

| To get a whole number divisor, multiply both .03 and 4.8 by 100. | → | $$\begin{array}{r}160\\3\overline{)480}\\300\\\hline 180\\180\\\hline 0\\0\\\hline 0\end{array}$$ | $$\begin{array}{r}160\\.03\overline{)4.80}\\300\\\hline 180\\180\\\hline 0\\0\\\hline 0\end{array}$$ |

Check: $160 \times .03 =$ _____

Divide.

	a	*b*	*c*	*d*
1.	$.8\overline{)9.6}$	$.3\overline{)\,.2\,8\,2}$	$.6\overline{)2.1\,6}$	$.4\overline{)\,.0\,1\,3\,6}$
2.	$.4\overline{)7\,2}$	$.08\overline{)1\,1\,2}$	$.03\overline{)3\,4\,5}$	$.002\overline{)1\,4}$
3.	$.03\overline{)\,.6}$	$.07\overline{)8.4}$	$.005\overline{)7\,5}$	$.9\overline{)2\,0.7}$

Perfect score: 12   My score: _____

Problem Solving

Solve each problem.

1. Joe divided 1.38 by .6 and Jim divided 1.38 by 2.3. What should Joe's answer be? What should Jim's answer be?

Joe's answer should be _____.

Jim's answer should be _____.

2. A stack of paper is 3.44 inches thick. How many stacks of paper each .8 inch thick could be formed? How thick would the partial stack be?

There could be _____ stacks formed.

The partial stack would be _____ inch thick.

3. Consider the numbers named by .032 and .08. What is the quotient if you divide the least number by the greatest number?

The quotient is _____.

4. In problem **3**, what is the quotient if you divide the greatest number by the least number?

The quotient is _____.

5. Arlene divided .692 by .4 and obtained an answer of .173. Trina divided .692 by .4 and obtained an answer of 17.3. Is either girl correct? What is the correct answer?

Is either girl correct? _____

The correct answer is _____.

6. Consider the numbers named by .004 and .2. What is the quotient if you divide the greatest number by the least number?

The quotient is _____.

7. In problem **6**, what is the quotient if you divide the least number by the greatest number?

The quotient is _____.

1.

2.

3.	4.

5.

6.	7.

Perfect score: 10   My score: _____

66

Lesson 9 Division

Study how to divide by a number like 1.2 or 2.05.

```
                              270
    1.2⟌324   →       1 2⟌3240
                              2400
                              ————
                               840
                               840
                               ———
                                 0
                                 0
                                 —
                                 0
```

```
                              4.1
  2.05⟌8.405   →     2 05⟌840.5
                             8200
                             ————
                              205
                              205
                              ———
                                0
```

Divide.

	a	*b*	*c*	*d*

1. .05⟌5 3.5 .16⟌2.0 8 2.5⟌.0 6 2 5 3.2⟌4 8

2. .06⟌1.6 2 1.7⟌2.2 1 2.3⟌1 2 8 8 .46⟌1 6 1

3. 1.03⟌.9 2 7 1.4⟌.2 5 2 .034⟌1 4.6 2 3.06⟌3 0.6

4. .008⟌2.7 2 .43⟌5 1.6 2.21⟌1 3.9 2 3 .016⟌2 4

Perfect score: 16   My score: _____

Problem Solving

Solve each problem.

1. A person needs about 2.5 quarts of water each day to survive. For how many days could a person survive on 40 quarts of water?

A person could survive for _____ days.

2. 2.6 pounds of meat costs $7.80. What is the price per pound?

The price per pound is $_____.

3. Last week Felipe worked 3.5 hours and earned $18.41. How much money did he earn per hour?

He earned $_____ per hour.

4. Barb multiplied a number by .016. She obtained the answer .0192. What was the number?

The number was _____.

5. Mr. Mills paid $16.66 for 3.4 kilograms of grass seed. How much did the seed cost per kilogram?

The seed cost $_____ per kilogram.

6. Heather paid $9.28 for gasoline at a service station. The gasoline cost $.29 per liter. How many liters did she buy?

She bought _____ liters.

7. The odometer shows how many miles Darlene rode her bicycle in .75 hour. What was her average speed?

`0 0 0 8 . 4`

Her average speed was _____ miles per hour.

8. Consider the numbers named by .18, 1.8, and .018. What is the quotient if you divide the greatest number by the least number?

The quotient is _____.

1.	2.
3.	4.
5.	6.
7.	8.

Perfect score: 8   My score: _____

68

Lesson 10   Mixed Practice

Add or subtract.

1. $31.9 + 42.8 + 61.3 =$ _____

2. $3.72 - 1.4 =$ _____

3. $32.91 + .47 + 1.6 =$ _____

4. $5.4 - 2.45 =$ _____

5. $751 - 7.51 =$ _____

6. $423 + .423 + 4.23 =$ _____

7. $3.18 - 2.486 =$ _____

Multiply or divide.

8. $6 \times 24.1 =$ _____

9. $44.1 \div 7 =$ _____

10. $.007 \times 3.2 =$ _____

11. $7.62 \div .06 =$ _____

12. $4.21 \times 4.21 =$ _____

13. $4.92 \div .012 =$ _____

14. $1.3 \times .4163 =$ _____

Perfect score: 14   My score: _____

Problem Solving

Solve each problem.

1. Consider the numbers named by .034; 2.35; 3; .74; and 5.95. What is the difference between the two least numbers?

The difference is _____.

2. In problem **1**, what is the product of the two greatest numbers?

The product is _____.

3. In problem **1**, what is the quotient if you divide the greatest number by the least number?

The quotient is _____.

4. In problem **1**, what is the sum of all five of the numbers?

The sum is _____.

5. Cherry multiplied some number by 4.8 and got the product .624. What was the number?

The number was _____.

6. A bar 3.450 centimeters wide is .065 centimeter wider than it should be. How wide should the bar be?

The bar should be _____ centimeters wide.

7. At $5.35 per kilogram, how much would it cost for a 1.6-kilogram roast?

It would cost $_____.

8. Gus paid $8.58 for 7.8 gallons of gasoline. What was the price per gallon?

The price was $_____ per gallon.

9. Babe Ruth's lifetime batting average was .342. Lou Gehrig's lifetime batting average was .340. Which player had the higher average? How much higher was his average?

_____ had the higher average.

His average was _____ higher.

1.	2.
3.	**4.**
5.	**6.**
7.	**8.**
9.	

Perfect score: 10   My score: _____

70

CHAPTER 3 TEST

Add or subtract.

	a	*b*	*c*	*d*
1.	.4 7 +1.8 3	1 2.9 6 5 4.7 8 2 +3.9 8 6	4.0 1 2 7.1 +8 3	.0 6 7 1.3 8 +1 6.4 2 5

2.	3 4.1 8 2 −1 6.4 5 3	4.2 1 −.8 4	4.1 −1.6 8	2 7 −1.2 8 9

Multiply or divide.

3.	1 7.3 ×.5	.6 8 ×.3	.7 2 ×.0 9	.1 6 ×7

4. 6)8.4   9).1 2 7 8   .5).1 5 6 5   .07)2 9 4

5.	4 2.1 3 ×.0 3 4	7 2.9 ×8.3 6	2.6)1 2.2 2	.32).1 3 1 2

Perfect score: 20   My score: _____

PRE-TEST—Ratio and Proportion

Write each of the following as a ratio in two ways as shown.

	a	**b**
1. 8 runs in 6 innings	8 to 6	$\frac{8}{6}$
2. 2 pies for 10 persons	_____	_____
3. 8 books to 2 tables	_____	_____
4. 3 assignments in 5 days	_____	_____

Draw a ring around each pair of ratios that are equal.

	a	*b*
5.	$\frac{5}{8}, \frac{15}{24}$	$\frac{7}{12}, \frac{11}{18}$
6.	$\frac{3}{4}, \frac{16}{20}$	$\frac{2}{3}, \frac{32}{48}$
7.	$\frac{4}{5}, \frac{24}{30}$	$\frac{5}{6}, \frac{20}{24}$

Solve each of the following.

	a	*b*
8.	$\frac{7}{9} = \frac{21}{n}$	$\frac{2}{3} = \frac{n}{24}$
9.	$\frac{n}{13} = \frac{35}{65}$	$\frac{8}{n} = \frac{4}{3}$
10.	$\frac{3}{5} = \frac{n}{100}$	$\frac{n}{7} = \frac{20}{28}$

Perfect score: 18   My score: _____

Lesson 1   Ratio

A **ratio** is a comparison of the numbers of two sets.

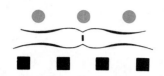

The ratio of ●'s to ■'s

is 3 to 4   or $\frac{3}{4}$.

The ratio of ■'s to ●'s

is 4 to 3   or $\frac{4}{3}$.

The ratio of ★'s to ⬡'s

is 3 to 2   or $\frac{3}{2}$.

The ratio of ⬡'s to ★'s

is 2 to 3   or $\frac{2}{3}$.

Write each of the following as a ratio in two ways as shown.

		a	*c*
1.	1 pen to 4 pencils	1 to 4	$\frac{1}{4}$
2.	9 chairs to 2 tables		
3.	3 books for 6 students		
4.	8 hits for 9 innings		
5.	3 teachers for 60 students		
6.	15 candy bars for 5 boys		
7.	6 miles in 2 hours		
8.	6 pages in 30 minutes		
9.	3 quarts for 1 dollar		
10.	4 dollars for 3 children		

Perfect score: 18   My score: _____

Lesson 2 Proportions

A **proportion** expresses the equality of two ratios.

$$\frac{2}{3}=\frac{4}{6} \qquad\qquad \frac{1}{5}=\frac{4}{20} \qquad\qquad \frac{3}{4}=\frac{9}{12} \qquad\qquad \frac{n}{6}=\frac{10}{12}$$

Study how you can tell if two ratios are equal.

$\frac{1}{4}\diagdown\frac{2}{8} \;\dashrightarrow 4\times2=8$
$\phantom{\frac{1}{4}}\dashrightarrow 1\times8=8$

$\frac{1}{4}=\frac{2}{8}$ since $\underline{\;\;1\;\;\times\;\;8\;\;}=\underline{\;\;4\;\;\times\;\;2\;\;}$.

$\frac{3}{5}\diagdown\frac{6}{10}\dashrightarrow 5\times6=30$
$\phantom{\frac{3}{5}}\dashrightarrow 3\times10=30$

$\frac{3}{5}=\frac{6}{10}$ since $\underline{\;\;\;\;\times\;\;\;\;}=\underline{\;\;\;\;\times\;\;\;\;}$.

Draw a ring around each pair of ratios that are equal.

	a	*b*
1.	$\frac{2}{3}, \frac{6}{9}$	$\frac{7}{8}, \frac{13}{16}$
2.	$\frac{5}{16}, \frac{11}{32}$	$\frac{4}{5}, \frac{8}{10}$
3.	$\frac{5}{8}, \frac{10}{16}$	$\frac{3}{4}, \frac{16}{20}$
4.	$\frac{1}{6}, \frac{2}{12}$	$\frac{3}{7}, \frac{9}{14}$
5.	$\frac{3}{9}, \frac{1}{3}$	$\frac{3}{2}, \frac{18}{12}$
6.	$\frac{5}{6}, \frac{9}{12}$	$\frac{6}{16}, \frac{3}{8}$
7.	$\frac{16}{20}, \frac{9}{10}$	$\frac{15}{24}, \frac{5}{8}$
8.	$\frac{10}{12}, \frac{4}{5}$	$\frac{12}{16}, \frac{3}{4}$
9.	$\frac{7}{25}, \frac{28}{100}$	$\frac{9}{20}, \frac{48}{100}$
10.	$\frac{5}{16}, \frac{4}{12}$	$\frac{7}{8}, \frac{21}{24}$

Perfect score: 20   My score: _____

Lesson 3   Proportions

Study how a proportion can be used in problem solving.

The ratio of the number of boys in the choir to the number of girls in the choir is 6 to 7. If there are 24 boys in the choir, how many girls are in the choir?

$\dfrac{6}{7} = \dfrac{24}{n}$ — There are 24 boys in the choir.

— Let n represent the number of girls in the choir.

$6 \times n = 7 \times 24$ What must $6 \times n$ be divided by to obtain n? _____

$\dfrac{6 \times n}{6} = \dfrac{7 \times 24}{6}$ Since $6 \times n$ is divided by 6, 7×24 must also be divided by _____.

$\dfrac{\overset{1}{\cancel{6}} \times n}{\underset{1}{\cancel{6}}} = \dfrac{7 \times \overset{4}{\cancel{24}}}{\underset{1}{\cancel{6}}}$

$n = 7 \times 4$ or 28   There are _____ girls in the choir.

Solve each of the following.

	a	b	c
1.	$\dfrac{7}{8} = \dfrac{14}{n}$	$\dfrac{5}{24} = \dfrac{10}{n}$	$\dfrac{7}{12} = \dfrac{21}{n}$
2.	$\dfrac{10}{6} = \dfrac{5}{n}$	$\dfrac{35}{40} = \dfrac{7}{n}$	$\dfrac{9}{10} = \dfrac{90}{n}$
3.	$\dfrac{7}{9} = \dfrac{35}{n}$	$\dfrac{4}{6} = \dfrac{2}{n}$	$\dfrac{27}{36} = \dfrac{3}{n}$
4.	$\dfrac{4}{16} = \dfrac{2}{n}$	$\dfrac{17}{51} = \dfrac{1}{n}$	$\dfrac{2}{14} = \dfrac{10}{n}$

Perfect score: 12   My score: _____

Problem Solving

Use a proportion to solve each problem.

1. A baseball team has won 14 games. The ratio of games won to games played is 7 to 12. How many games has the team played?

The team has played _____ games.

2. A cleaning concentrate is to be mixed with water so that the ratio of water to concentrate is 8 to 1. At that rate how much concentrate should be mixed with 16 quarts of water?

_____ quarts of concentrate should be mixed with 16 quarts of water.

3. The ratio of boys to girls at Lincoln School is 11 to 10. There are 330 boys at the school. How many girls are there?

There are _____ girls at the school.

4. At the cafeteria the ratio of pints of chocolate milk sold to pints of plain milk sold is 4 to 7. At this rate how many pints of plain milk will be sold if 100 pints of chocolate milk are sold?

_____ pints of plain milk will be sold.

5. During a Cougars' basketball game, the ratio of baskets made to baskets attempted was 2 to 5. The Cougars made 26 baskets. How many baskets did they attempt?

They attempted _____ baskets.

6. The park department planted oak and maple trees in a 5 to 7 ratio. They planted 105 oaks. How many maples did they plant?

They planted _____ maples.

7. On a spelling test, the ratio of words Dawn spelled correctly to those she misspelled was 4 to 1. She spelled 20 words correctly. How many did she misspell?

She misspelled _____ words.

1.	
2.	3.
4.	5.
6.	7.

Perfect score: 7   My score: _____

76

Lesson 4 Proportions

Study how the proportions below are solved.

$\dfrac{7}{8}=\dfrac{21}{n}$

$7\times n=8\times 21$

$\dfrac{7\times n}{7}=\dfrac{8\times 21}{7}$

$\dfrac{\overset{1}{\cancel{7}}\times n}{\underset{1}{\cancel{7}}}=\dfrac{8\times \overset{3}{\cancel{21}}}{\underset{1}{\cancel{7}}}$

$n=8\times 3$

$n=\underline{\ \ 24\ \ }$

$\dfrac{2}{3}=\dfrac{n}{12}$

$2\times 12=3\times n$

$\dfrac{2\times 12}{3}=\dfrac{3\times n}{3}$

$\dfrac{2\times \overset{4}{\cancel{12}}}{\underset{1}{\cancel{3}}}=\dfrac{\overset{1}{\cancel{3}}\times n}{\underset{1}{\cancel{3}}}$

$2\times 4=n$

$\underline{\ \ \ \ \ \ }=n$

$\dfrac{1}{n}=\dfrac{5}{20}$

$1\times 20=n\times 5$

$\dfrac{1\times 20}{5}=\dfrac{n\times 5}{5}$

$\dfrac{1\times \overset{4}{\cancel{20}}}{\underset{1}{\cancel{5}}}=\dfrac{n\times \overset{1}{\cancel{5}}}{\underset{1}{\cancel{5}}}$

$1\times 4=n$

$\underline{\ \ \ \ \ \ }=n$

$\dfrac{n}{2}=\dfrac{12}{6}$

$n\times 6=2\times 12$

$\dfrac{n\times 6}{6}=\dfrac{2\times 12}{6}$

$\dfrac{n\times \overset{1}{\cancel{6}}}{\underset{1}{\cancel{6}}}=\dfrac{2\times \overset{2}{\cancel{12}}}{\underset{1}{\cancel{6}}}$

$n=2\times 2$

$n=\underline{\ \ \ \ \ \ }$

Solve each of the following.

	a	*b*	*c*
1.	$\dfrac{3}{8}=\dfrac{6}{n}$	$\dfrac{7}{9}=\dfrac{n}{27}$	$\dfrac{n}{30}=\dfrac{5}{6}$
2.	$\dfrac{5}{n}=\dfrac{35}{49}$	$\dfrac{3}{4}=\dfrac{n}{100}$	$\dfrac{n}{32}=\dfrac{3}{8}$
3.	$\dfrac{n}{6}=\dfrac{6}{9}$	$\dfrac{10}{n}=\dfrac{5}{8}$	$\dfrac{5}{3}=\dfrac{10}{n}$
4.	$\dfrac{8}{11}=\dfrac{n}{99}$	$\dfrac{7}{8}=\dfrac{n}{1000}$	$\dfrac{n}{56}=\dfrac{7}{8}$

Perfect score: 12   My score: _____

Proportions

A certain machine can produce 7 parts in 2 hours. At that rate, how many parts can be produced in 8 hours?

Let n stand for the number of parts made in 8 hours.

You can compare the number of parts to the number of hours.	or	You can compare the number of hours to the number of parts.

parts $----\rightarrow$ hours $----\rightarrow$
$$\frac{7}{2} = \frac{n}{8}$$
$$7 \times 8 = 2 \times n$$
$$\frac{7 \times 8}{2} = \frac{2 \times n}{2}$$
$$\underline{\hspace{1cm}} = n$$

hours $----\rightarrow$ parts $----\rightarrow$
$$\frac{2}{7} = \frac{8}{n}$$
$$2 \times n = 7 \times 8$$
$$\frac{2 \times n}{2} = \frac{7 \times 8}{2}$$
$$n = \underline{\hspace{1cm}}$$

In 8 hours the machine can make _____ parts.

Use a proportion to solve each problem.

1. In 3 hours, 225 cases of corn can be canned. At that rate, how many cases of corn can be canned in 8 hours?

_____ cases of corn can be canned.

2. A plane traveled 1,900 kilometers in 2 hours. At that rate, how far will the plane travel in 3 hours?

The plane will travel _____ kilometers.

3. A duplicating machine can produce 50 copies in 2 minutes. At that rate, how long will it take to produce 150 copies?

It will take _____ minutes.

4. Six cases of merchandise weigh 180 pounds. How much would 10 such cases weigh?

10 cases weigh _____ pounds.

1.

2.

3.

4.

Perfect score: 4   My score: _____

78

Lesson 5   Problem Solving

NAME _____

Use a proportion to solve each problem.

1. Mrs. Fisher wants to buy 12 pounds of apples. What would be the cost if she buys the kind of apples shown above?

The cost would be $_____.

2. What would be the cost of 3 bunches of the green onions shown above?

The cost would be $_____.

3. What would be the cost of 4 pounds of the carrots shown above?

The cost would be $_____.

4. What would be the cost of 6 pounds of the navel oranges shown above?

The cost would be $_____.

5. Mr. Sims needs 3 pounds of asparagus. What would be the cost of the asparagus?

The cost would be $_____.

1.
2.
3.
4.
5.

Perfect score: 5   My score: _____

Problem Solving

Use a proportion to solve each problem.

1. David won an election by a 5 to 2 margin over Estelle. David received 35 votes. How many votes did Estelle receive?

Estelle received _____ votes.

2. The sales tax on a $100 purchase is $5. How much would the sales tax be on a $40 purchase?

The sales tax would be $_____.

3. On a map each inch represents 60 miles. What is the distance between two towns which are represented 5 inches apart on the map?

The distance is _____ miles.

4. Three bushels of apples weigh 105 pounds. How much would 2 bushels of apples weigh?

Two bushels would weigh _____ pounds.

5. To make a certain grade of concrete, 2 bags of cement are mixed with 300 pounds of sand. How much sand should be mixed with 12 bags of cement?

_____ pounds of sand should be mixed with 12 bags of cement.

6. A baseball player is hitting 2 home runs in every 25 times at bat. Suppose he continues at that rate. How many home runs will he hit in 550 times at bat?

He will hit _____ home runs.

7. A 3-inch by 5-inch photograph is enlarged so that the longer side is 15 inches. What is the length of the shorter side of the enlarged photograph?

It is _____ inches long.

| 1. |
| 2. |
| 3. |
| 4. |
| 5. |
| 6. |
| 7. |

Perfect score: 7   My score: _____

80

Lesson 6   Proportions

Solve each of the following.

	a	b	c
1.	$\frac{n}{2} = \frac{5}{10}$	$\frac{n}{1} = \frac{28}{7}$	$\frac{n}{15} = \frac{4}{12}$
2.	$\frac{3}{7} = \frac{n}{56}$	$\frac{75}{100} = \frac{n}{8}$	$\frac{26}{16} = \frac{n}{8}$
3.	$\frac{4}{n} = \frac{24}{36}$	$\frac{15}{n} = \frac{35}{28}$	$\frac{14}{n} = \frac{7}{13}$
4.	$\frac{1}{6} = \frac{16}{n}$	$\frac{15}{75} = \frac{4}{n}$	$\frac{21}{49} = \frac{3}{n}$
5.	$\frac{n}{15} = \frac{64}{12}$	$\frac{11}{24} = \frac{n}{72}$	$\frac{8}{n} = \frac{12}{18}$
6.	$\frac{6}{14} = \frac{15}{n}$	$\frac{n}{6} = \frac{20}{8}$	$\frac{8}{25} = \frac{n}{100}$
7.	$\frac{5}{n} = \frac{20}{24}$	$\frac{3}{8} = \frac{9}{n}$	$\frac{3}{5} = \frac{n}{100}$

Perfect score: 21   My score: _____

Problem Solving

Use a proportion to solve each problem.

1. A 2-kilogram roast requires 90 minutes to cook. At that rate, how long should a 3-kilogram roast be cooked?

It should be cooked _____ minutes.

2. A jogger burned 96 calories during 8 minutes of jogging. At that rate, how many calories would be burned during 14 minutes of jogging?

_____ calories would be burned.

3. On a certain map 2 inches represent 250 miles. What distance would 3 inches represent on this map?

Three inches would represent _____ miles.

4. Six tons of ore can be processed in 3 hours. At that rate how many tons of ore can be processed in 8 hours?

_____ tons of ore can be processed.

5. An automobile used 8 gallons of gasoline in traveling 120 miles. At that rate, how many gallons will be used in traveling 360 miles?

_____ gallons of gasoline will be used.

6. Nine items can be produced in 15 minutes. How many items can be produced in 35 minutes?

_____ items can be produced in 35 minutes.

7. A baseball player has hit 9 home runs in 54 games. At that rate how many home runs will he hit during a 162-game season?

He will hit _____ home runs.

8. A 9-inch piece of rubber can be stretched to a length of 15 inches. At that rate, to what length can a 12-inch piece of rubber be stretched?

It can be stretched to a length of _____ inches.

1.

2.

3.

4.

5.

6.

7.

8.

Perfect score: 8   My score: _____

82

CHAPTER 4 TEST

NAME _____

Write each of the following as a ratio in two ways.

	a	b

1. 28 people for 3 elevators

2. 5 women to 2 men

3. 17 gallons in 200 miles

4. 4 books for 7 students

Solve each of the following.

	a	b	c
5.	$\frac{5}{8}=\frac{20}{n}$	$\frac{1}{3}=\frac{n}{42}$	$\frac{n}{5}=\frac{8}{10}$
6.	$\frac{5}{n}=\frac{20}{48}$	$\frac{1}{2}=\frac{16}{n}$	$\frac{3}{4}=\frac{n}{1000}$
7.	$\frac{n}{50}=\frac{54}{100}$	$\frac{7}{n}=\frac{35}{100}$	$\frac{18}{25}=\frac{n}{100}$

Use a proportion to solve each problem.

8. A runner ran 2 miles in 10 minutes. At that rate, how long will it take to run 6 miles?

It will take _____ minutes.

8.

9. A car traveled 237 kilometers in 3 hours. At that rate how far could the car travel in 4 hours? How far could the car travel in 5 hours?

_____ kilometers could be traveled in 4 hours.

_____ kilometers could be traveled in 5 hours.

9.

Perfect score: 20   My score: _____

PRE-TEST—Decimals, Fractions, and Percent

Complete the following.

	a		b

1. $\dfrac{7}{10} =$ _____% $\dfrac{3}{4} =$ _____%

2. $.9 =$ _____% $.14 =$ _____%

3. $1.3 =$ _____% $\dfrac{3}{8} =$ _____%

4. $.725 =$ _____% $1.672 =$ _____%

Change each of the following to a decimal.

5. $7\% =$ _____ $69\% =$ _____

6. $12.5\% =$ _____ $3.25\% =$ _____

7. $18.76\% =$ _____ $7.125\% =$ _____

8. $7\dfrac{3}{4}\% =$ _____ $9\dfrac{1}{4}\% =$ _____

Change each of the following to a fraction or mixed numeral in simplest form.

9. $27\% =$ _____ $35\% =$ _____

10. $117\% =$ _____ $280\% =$ _____

11. $54\% =$ _____ $175\% =$ _____

12. $3\% =$ _____ $5\% =$ _____

Perfect score: 24   My score: _____

Lesson 1   Percent

A symbol such as 3% (read 3 percent), 17%, 125%, and so on, expresses the ratio of some number to 100.

$5\% = \frac{5}{100}$ or .05 $29\% = \frac{29}{100}$ or .29 $123\% = \frac{123}{100}$ or 1.23

$7\% = $_____ or _____ $57\% = $_____ or _____ $159\% = $_____ or _____

Complete the following.

	percent	fraction	decimal
1.	1%	_____	_____
2.	21%	_____	_____
3.	129%	_____	_____
4.	3%	_____	_____
5.	39%	_____	_____
6.	143%	_____	_____
7.	9%	_____	_____
8.	51%	_____	_____
9.	169%	_____	_____
10.	69%	_____	_____
11.	233%	_____	_____
12.	83%	_____	_____
13.	357%	_____	_____
14.	99%	_____	_____

Perfect score: 28   My score: _____

Lesson 2 Fractions and Percent

Study how a proportion is used to change a fraction to a percent.

Change $\frac{3}{4}$ to a percent.

$$\frac{3}{4} = \frac{n}{100}$$

$$3 \times 100 = 4 \times n$$

$$\frac{3 \times 100}{4} = \frac{4 \times n}{4}$$

$$75 = n$$

$$\frac{3}{4} = \frac{75}{100} \text{ or } \underline{\hspace{1cm} 75 \hspace{1cm}}\%.$$

Change $\frac{5}{8}$ to a percent.

$$\frac{5}{8} = \frac{n}{100}$$

$$5 \times 100 = 8 \times n$$

$$\frac{5 \times 100}{8} = \frac{8 \times n}{8}$$

$$\frac{500}{8} = 62\frac{1}{2} \rightarrow 62\frac{1}{2} = n$$

$$\frac{5}{8} = \frac{62\frac{1}{2}}{100} \text{ or } \underline{\hspace{1cm} 62\frac{1}{2} \hspace{1cm}}\%$$

Complete the following.

	a		b

1. $\frac{1}{2} = $ _____ % $\frac{1}{4} = $ _____ %

2. $\frac{1}{10} = $ _____ % $\frac{1}{8} = $ _____ %

3. $\frac{1}{5} = $ _____ % $\frac{3}{8} = $ _____ %

4. $\frac{2}{5} = $ _____ % $\frac{7}{20} = $ _____ %

5. $\frac{3}{10} = $ _____ % $\frac{7}{8} = $ _____ %

6. $\frac{4}{5} = $ _____ % $\frac{6}{25} = $ _____ %

7. $\frac{1}{3} = $ _____ % $\frac{7}{10} = $ _____ %

8. $\frac{2}{3} = $ _____ % $\frac{49}{50} = $ _____ %

Perfect score: 16   My score: _____

Lesson 3   Fractions and Percent

Study how a percent is changed to a fraction or a mixed numeral in simplest form.

$$65\% = \frac{65}{100}$$
$$= \frac{13}{20}$$

$$125\% = \frac{125}{100}$$
$$= \frac{5}{4} \text{ or } 1\frac{1}{4}$$

$$16\% = \frac{16}{100}$$
$$= \underline{\qquad}$$

Change each of the following to a fraction or a mixed numeral in simplest form.

a   b

1.   25% = _____   50% = _____

2.   30% = _____   75% = _____

3.   8% = _____   80% = _____

4.   60% = _____   10% = _____

5.   5% = _____   40% = _____

6.   20% = _____   56% = _____

7.   12% = _____   95% = _____

8.   100% = _____   150% = _____

9.   175% = _____   180% = _____

10.   78% = _____   85% = _____

11.   64% = _____   190% = _____

12.   350% = _____   2% = _____

Perfect score: 24   My score: _____

Fractions and Percent

Complete the following.

          a   *b*

1. $\dfrac{3}{5} = $ _____% $\dfrac{5}{8} = $ _____%

2. $\dfrac{3}{4} = $ _____% $\dfrac{9}{10} = $ _____%

3. $\dfrac{4}{25} = $ _____% $\dfrac{17}{50} = $ _____%

4. $\dfrac{2}{3} = $ _____% $\dfrac{11}{20} = $ _____%

Change each of the following to a fraction or mixed numeral in simplest form.

          a   *b*

5.   75% = _____   90% = _____

6.   250% = _____   28% = _____

7.   78% = _____   140% = _____

8.   6% = _____   62% = _____

9.   45% = _____   66% = _____

10.   82% = _____   26% = _____

11.   120% = _____   18% = _____

12.   4% = _____   325% = _____

Perfect score: 24   My score: _____

Lesson 4   Decimals and Percent

Study how a decimal is changed to a percent.

$.3 = .30$
$= \frac{30}{100}$
$= \underline{\quad 30 \quad}\%$

$.35 = \frac{35}{100}$
$= \underline{\qquad}\%$

$1.83 = \frac{183}{100}$
$= \underline{\qquad}\%$

$.335 = .335 \times 1$
$= \frac{.335}{1} \times \frac{100}{100}$
$= \frac{33.5}{100}$
$= \underline{\quad 33.5 \quad}\%$

Complete the following.

	a	b	c
1.	$.7 = \underline{\qquad}\%$	$.9 = \underline{\qquad}\%$	$.5 = \underline{\qquad}\%$
2.	$1.3 = \underline{\qquad}\%$	$2.6 = \underline{\qquad}\%$	$1.9 = \underline{\qquad}\%$
3.	$.03 = \underline{\qquad}\%$	$.84 = \underline{\qquad}\%$	$.35 = \underline{\qquad}\%$
4.	$1.06 = \underline{\qquad}\%$	$1.87 = \underline{\qquad}\%$	$2.45 = \underline{\qquad}\%$
5.	$.015 = \underline{\qquad}\%$	$.225 = \underline{\qquad}\%$	$.375 = \underline{\qquad}\%$
6.	$1.032 = \underline{\qquad}\%$	$2.125 = \underline{\qquad}\%$	$1.477 = \underline{\qquad}\%$
7.	$.08 = \underline{\qquad}\%$	$.875 = \underline{\qquad}\%$	$3.75 = \underline{\qquad}\%$
8.	$1.24 = \underline{\qquad}\%$	$1.3 = \underline{\qquad}\%$	$2.001 = \underline{\qquad}\%$
9.	$.1 = \underline{\qquad}\%$	$.396 = \underline{\qquad}\%$	$.17 = \underline{\qquad}\%$
10.	$1.296 = \underline{\qquad}\%$	$2.39 = \underline{\qquad}\%$	$2.3 = \underline{\qquad}\%$

Perfect score: 30   My score: _____

Lesson 5   Decimals and Percent

Study how a percent is changed to a decimal.

Shortcut

Remove the % and move the decimal point 2 places to the left.

$21.5\% = \dfrac{21.5}{100}$

$= 21.5 \times \dfrac{1}{100}$

$= 21.5 \times .01$

$= \underline{\quad .215 \quad}$

21.5%   21.5%   .215

$47\% = \dfrac{47}{100}$

$= \underline{\quad .47 \quad}$

47%   47%   _____

$8\% = \dfrac{8}{100}$

$= \underline{\qquad}$

8%   08%   _____

Change each of the following to a decimal.

	a	b	c
1.	85% = _____	12% = _____	35% = _____
2.	4% = _____	6% = _____	7% = _____
3.	25% = _____	75% = _____	33% = _____
4.	99.4% = _____	56.25% = _____	18.75% = _____
5.	100% = _____	62.5% = _____	87.5% = _____
6.	16.4% = _____	93.75% = _____	250% = _____
7.	9.6% = _____	6.25% = _____	475% = _____
8.	54% = _____	125% = _____	19.5% = _____

Perfect score: 24   My score: _____

Lesson 6   Decimals and Percent

Change $4\frac{1}{2}\%$ to a decimal.
Since $\frac{1}{2} = .5,$

$$4\frac{1}{2}\% = \underline{\quad 4.5 \quad}\%.$$

04.5% ⟶ .045

$$4\frac{1}{2}\% = \underline{\quad .045 \quad}$$

Change $6\frac{3}{4}\%$ to a decimal.
Since $\frac{3}{4} = .75,$

$$6\frac{3}{4}\% = \underline{\qquad\qquad}\%.$$

06.75% ⟶ .0675

$$6\frac{3}{4}\% = \underline{\qquad\qquad}$$

Change each of the following to a decimal.

	a	b	c
1.	$8\frac{1}{2}\% = \underline{\qquad}$	$7\frac{3}{4}\% = \underline{\qquad}$	$6\frac{2}{5}\% = \underline{\qquad}$
2.	$5\frac{1}{4}\% = \underline{\qquad}$	$12\frac{1}{2}\% = \underline{\qquad}$	$8\frac{1}{10}\% = \underline{\qquad}$
3.	$18\frac{3}{4}\% = \underline{\qquad}$	$6\frac{1}{4}\% = \underline{\qquad}$	$7\frac{4}{5}\% = \underline{\qquad}$
4.	$37\frac{1}{2}\% = \underline{\qquad}$	$9\frac{3}{4}\% = \underline{\qquad}$	$5\frac{1}{5}\% = \underline{\qquad}$
5.	$8\frac{1}{4}\% = \underline{\qquad}$	$87\frac{1}{2}\% = \underline{\qquad}$	$9\frac{3}{5}\% = \underline{\qquad}$
6.	$68\frac{3}{4}\% = \underline{\qquad}$	$7\frac{3}{10}\% = \underline{\qquad}$	$7\frac{1}{2}\% = \underline{\qquad}$
7.	$8\frac{2}{5}\% = \underline{\qquad}$	$16\frac{1}{4}\% = \underline{\qquad}$	$9\frac{7}{10}\% = \underline{\qquad}$
8.	$6\frac{1}{2}\% = \underline{\qquad}$	$93\frac{3}{4}\% = \underline{\qquad}$	$8\frac{1}{5}\% = \underline{\qquad}$
9.	$37\frac{2}{5}\% = \underline{\qquad}$	$3\frac{1}{2}\% = \underline{\qquad}$	$31\frac{1}{4}\% = \underline{\qquad}$
10.	$83\frac{9}{10}\% = \underline{\qquad}$	$42\frac{3}{5}\% = \underline{\qquad}$	$21\frac{3}{4}\% = \underline{\qquad}$
11.	$2\frac{7}{50}\% = \underline{\qquad}$	$12\frac{1}{4}\% = \underline{\qquad}$	$1\frac{2}{5}\% = \underline{\qquad}$

Perfect score: 33   My score: _____

Decimals and Percent

Complete the following.

	a decimal	a percent		b percent	b decimal
1.	.5	_____		7.4%	_____
2.	1.7	_____		16.8%	_____
3.	2.8	_____		137.5%	_____
4.	.09	_____		7.49%	_____
5.	.67	_____		86.67%	_____
6.	1.15	_____		209.86%	_____
7.	.075	_____		7.125%	_____
8.	.375	_____		3.75%	_____
9.	.007	_____		$5\frac{1}{2}\%$	_____
10.	1.414	_____		$7\frac{1}{4}\%$	_____
11.	.0875	_____		$4\frac{3}{4}\%$	_____
12.	.0095	_____		$93\frac{3}{4}\%$	_____
13.	1.0655	_____		$62\frac{1}{2}\%$	_____
14.	.0009	_____		$35\frac{2}{5}\%$	_____
15.	.3125	_____		$9\frac{3}{10}\%$	_____

Perfect score: 30   My score: _____

Lesson 7   Percent

Complete the following. Express each fraction in simplest form.

	a			b	
	percent	fraction		percent	decimal
1.	5%	_____		8%	_____
2.	_____	$\frac{1}{10}$		_____	.09
3.	12%	_____		12%	_____
4.	_____	$\frac{1}{8}$		_____	.27
5.	15%	_____		7.5%	_____
6.	_____	$\frac{1}{5}$		_____	.168
7.	25%	_____		125%	_____
8.	_____	$\frac{3}{4}$		_____	1.35
9.	37%	_____		$6\frac{1}{2}\%$	_____
10.	_____	$\frac{2}{5}$		_____	.2748
11.	50%	_____		206.5%	_____
12.	_____	$\frac{3}{5}$		_____	1.375
13.	62%	_____		$8\frac{1}{4}\%$	_____
14.	_____	$\frac{7}{8}$		_____	.0675
15.	83%	_____		$7\frac{3}{4}\%$	_____

Perfect score: 30   My score: _____

Problem Solving

Solve each problem.

1. By what percent have the regular prices of suits been reduced?

The prices have been reduced _____%.

2. Suppose the sales tax on a suit is 6½%. Write the percent of sales tax as a decimal.

The percent of sales tax written as a decimal is _____.

3. Suppose the store sells $\frac{1}{4}$ of its inventory of suits. What percent of the inventory of suits did it sell?

_____% of the inventory of suits were sold.

4. What fractional part of the regular price is the discount on suits?

The discount is _____ of the regular price.

5. What fractional part of the regular price is the sale price of each suit?

The sale price of each suit is _____ of the regular price.

1.	2.
3.	4.
5.	

Perfect score: 5   My score: _____

CHAPTER 5 TEST

Complete the following.

	percent	fraction	decimal
1.	9%	_____	_____
2.	47%	_____	_____
3.	79%	_____	_____
4.	153%	_____	_____

Complete the following.

$$a \qquad\qquad b \qquad\qquad c$$

5. $\dfrac{3}{10} =$ _____% $\dfrac{5}{8} =$ _____% $\dfrac{4}{5} =$ _____%

6. $1.2 =$ _____% $.325 =$ _____% $.3175 =$ _____%

Change each of the following to a decimal.

7. $62.25\% =$ _____ $126.5\% =$ _____ $8\dfrac{3}{4}\% =$ _____

Change each of the following to a fraction or mixed numeral in simplest form.

8. $85\% =$ _____ $160\% =$ _____ $6\% =$ _____

Perfect score: 20   My score: _____

PRE-TEST—Percent

Complete the following.

| *a* | *b* |

1. _____ is 40% of 80.   _____ is 35% of 200.

2. _____ is 50% of 23.   _____ is 25% of 63.

3. _____ is $3\frac{1}{2}$% of 120.   _____ is $62\frac{1}{2}$% of 240.

4. _____ is 7.4% of 800.   _____ is 9.6% of 700.

5. 35 is _____% of 140.   150 is _____% of 240.

6. 45 is _____% of 60.   108 is _____% of 90.

7. 4.6 is _____% of 9.2.   12.8 is _____% of 80.

8. .5 is _____% of 1.6.   .32 is _____% of 1.6.

9. 21 is 30% of _____.   49 is 35% of _____.

10. 42 is 30% of _____.   138 is 46% of _____.

11. 25 is 25% of _____.   .7 is 1% of _____.

12. .8 is 160% of _____.   33 is 25% of _____.

Perfect score: 24 My score: _____

Lesson 1 Percent of a Number

To find a percent of a number, you can name the percent as a fraction or as a decimal.

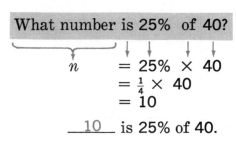

What number is 25% of 40?

$$n = 25\% \times 40$$
$$= \tfrac{1}{4} \times 40$$
$$= 10$$

___10___ is 25% of 40.

What number is $6\tfrac{1}{2}\%$ of 20?

$$n = 6\tfrac{1}{2}\% \times 20$$
$$= 6.5\% \times 20$$
$$= .065 \times 20$$
$$= 1.3$$

___1.3___ is $6\tfrac{1}{2}\%$ of 20.

Complete the following.

     a   *b*

1. _____ is 50% of 80.   _____ is 75% of 40.

2. _____ is 15% of 60.   _____ is 10% of 96.

3. _____ is 25% of 30.   _____ is 4% of 180.

4. _____ is 20% of 96.   _____ is 40% of 200.

5. _____ is 12% of 80.   _____ is 75% of 64.

6. _____ is 25% of 30.6.   _____ is 5% of 980.

7. _____ is 97% of 28.5.   _____ is 100% of 65.7.

8. _____ is $7\tfrac{1}{2}\%$ of 60.   _____ is $1\tfrac{1}{2}\%$ of 300.

9. _____ is $6\tfrac{1}{4}\%$ of 160.   _____ is $8\tfrac{3}{4}\%$ of 80.

Perfect score: 18 My score: _____

Problem Solving

Solve each problem.

1. Bob's team won 75% of their baseball games. They played 16 games in all. How many games did they win?

They won _____ games.

2. A factory is operating at 80% of capacity. The capacity is 300 cases per hour. How many cases are being produced each hour?

_____ cases are being produced each hour.

3. Of the 630 seats in the cafeteria, 70% are in use. How many seats are being used?

_____ seats are being used.

4. Of the 160 base hits a baseball player made last season, 35% were for extra bases. How many extra-base hits did the player make last season?

He made _____ extra-base hits.

5. Of the 240 trees the park department planted last month, 25% were maples. How many maple trees did they plant last month?

They planted _____ maple trees.

6. A salesman's commission is 6% of his total sales. His total sales last month were $24,000. How much was his commission last month?

His commission was $_____ last month.

7. After testing 7,200 transistors, an inspector found that 1.5% were defective. How many of the transistors were defective?

_____ transistors were defective.

1.	
2.	**3.**
4.	**5.**
6.	**7.**

Perfect score: 7   My score: _____

98

Lesson 2 Finding What Percent One Number Is of Another

3 is what percent of 25?

$$3 = \quad n\% \quad \times 25$$
$$3 = \frac{n}{100} \times 25$$
$$\frac{3}{1} = \frac{n \times 25}{100}$$
$$300 = n \times 25$$
$$12 = n$$

3 is __12__% of 25.

16.5 is what percent of 55?

$$16.5 = n\% \times 55$$
$$16.5 = \frac{n}{100} \times 55$$
$$\frac{16.5}{1} = \frac{n \times 55}{100}$$
$$1650 = n \times 55$$
$$30 = n$$

16.5 is __30__% of 55.

Complete the following.

a   *b*

1. 25 is _____% of 125.   75 is _____% of 200.

2. 13 is _____% of 52.   24 is _____% of 48.

3. 60 is _____% of 60.   63 is _____% of 70.

4. 6.3 is _____% of 75.   37.5 is _____% of 50.

5. 5 is _____% of 4.   350 is _____% of 200.

6. 12.5 is _____% of 100.   62.5 is _____% of 50.

7. 27.3 is _____% of 60.   15 is _____% of 75.

8. 120 is _____% of 60.   135 is _____% of 180.

9. 1.1 is _____% of 100.   90 is _____% of 60.

10. 36 is _____% of 60.   39 is _____% of 48.

Perfect score: 20   My score: _____

Problem Solving

Solve each problem.

1. Twenty-four of 30 workers take 1 hour for lunch. What percent of the workers take 1 hour for lunch?

_____% take 1 hour for lunch.

2. During a basketball game, Jamille attempted 15 baskets and made 6. What percent of the baskets that she attempted did she make?

She made _____% of the baskets attempted.

3. The gasoline tank on a lawn mower will hold 1 liter of gasoline. There is .2 liter of gasoline in the tank. The tank is what percent full?

The tank is _____% full.

4. There are 33 people at a meeting. Forty people had been asked to the meeting. What percent of the people asked to the meeting are at the meeting?

_____% of the people asked to the meeting are at the meeting.

5. During a sale the price of a pair of shoes was reduced $12. The shoes regularly sell for $48. The price reduction is what percent of the regular price?

The price reduction is _____% of the regular price.

6. The sales tax on a $2.80 purchase is $.14. The sales tax is what percent of the purchase price?

The sales tax is _____% of the purchase price.

7. Out of 2160 votes, Jack received 1350 votes. What percent of the votes did he receive?

He received _____% of the votes.

1.	
2.	3.
4.	5.
6.	7.

Perfect score: 7   My score: _____

100

Lesson 3   Finding a Number When a Percent of It Is Known

You can name a percent as a fraction or as a decimal.

105 is 35% of what number?	.6 is 15% of what number?

$$105 = 35\% \times \qquad n$$
$$105 = \frac{35}{100} \times n$$
$$\frac{105}{1} = \frac{35 \times n}{100}$$
$$10500 = 35 \times n$$
$$300 = n$$

105 is 35% of ___300___.

$$.6 = 15\% \times n$$
$$.6 = .15 \times n$$
$$\frac{.6}{.15} = \frac{.15 \times n}{.15}$$
$$4 = n$$

.6 is 15% of ___4___.

Complete the following.

	a	b
1.	15 is 25% of _____.	.8 is 40% of _____.
2.	9 is 50% of _____.	32 is 16% of _____.
3.	120 is 75% of _____.	185 is 25% of _____.
4.	7 is 140% of _____.	83 is 100% of _____.
5.	.8 is 25% of _____.	2.45 is 35% of _____.
6.	4 is 40% of _____.	.5 is 125% of _____.
7.	150 is 20% of _____.	1.28 is 25% of _____.
8.	360 is 150% of _____.	54 is 90% of _____.
9.	33 is 66% of _____.	80 is 100% of _____.
10.	8 is 8% of _____.	1.5 is 5% of _____.

Perfect score: 20   My score: _____

Problem Solving

Solve each problem.

1. In his last basketball game, Max made 6 baskets. This was 40% of the baskets that he attempted. How many baskets did he attempt during the game?

He attempted _____ baskets.

2. The down payment on a motor bike is $125. This is 25% of the selling price. How much does the motor bike sell for?

The motor bike costs $_____.

3. Last season 20% of the base hits made by a ball player were home runs. The player hit 42 home runs. How many base hits did he make in all?

He made _____ base hits in all.

4. At the beginning of each track practice, Steve runs $\frac{1}{2}$ mile. This is 40% of the running that he will do during practice. How far will he run during each practice?

He will run _____ miles during each practice.

5. During a sale the price of a pants suit was reduced $10. The reduction was 20% of the regular price. What was the regular price?

The regular price was $_____.

6. Yesterday only 13 students were absent from school. This was 4% of the enrollment. What is the school's enrollment?

_____ students are enrolled.

7. Mrs. Woods received a $.78 discount for paying a bill early. The discount was 5% of her bill. How much was the bill before discount?

The bill was $_____ before discount.

1.	
2.	**3.**
4.	**5.**
6.	**7.**

Perfect score: 7   My score: _____

102

Lesson 4   Percent

Complete the following.

	a		b

1. _____ is 60% of 300.   _____ is 65% of 780.

2. _____ is 125% of 64.   _____ is 150% of 160.

3. _____ is 7% of 560.   _____ is $2\frac{1}{2}$% of 96.

4. _____ is $8\frac{1}{4}$% of 76.   _____ is 8.5% of 7200.

5. 30 is _____% of 120.   12 is _____% of 16.

6. 225 is _____% of 90.   867 is _____% of 867.

7. 126 is _____% of 2100.   36 is _____% of 400.

8. 16 is _____% of 80.   225 is _____% of 900.

9. 56 is 40% of _____.   72 is 120% of _____.

10. 106 is 106% of _____.   975 is 125% of _____.

11. .5 is 50% of _____.   .75 is 20% of _____.

12. .6 is 150% of _____.   7.65 is 25% of _____.

Perfect score: 24   My score: _____

103

Problem Solving

Solve each problem.

1. To buy a new car, a down payment of 20% is required. If a car costs $12,000, what is the required down payment?

The required down payment is $_____.

2. You buy a car that costs $9,890. You pay 25% of the cost of the car as a down payment. How much did you pay for the down payment?

You paid $_____ for the down payment.

3. Mr. Gallegy is buying a car that sells for $14,000. He was allowed $4,900 for his present car. What percent of the price of the new car was he allowed for his present car?

He was allowed _____%.

4. Miss Lane made a down payment of $3,800 on a new car. That was 25% of the total price. What was the total price of the new car?

The total price was $_____.

5. Andy and Diane want to buy a new van. They are allowed $5,500 for their present car. They also made a down payment of $2,300. They still owe 40% of the regular price of the car. What is the regular price of the car?

The regular price of the car is $_____.

1.	
2.	**3.**
4.	**5.**

Perfect score: 5   My score: _____

104

Lesson 5   Percent

Complete the following.

	a	b

1.   _____ is 30% of 90.   34 is 17% of _____.

2.   42 is _____% of 56.   _____ is 5% of 63.

3.   16 is 40% of _____.   84 is _____% of 70.

4.   _____ is 65% of 35.   675 is 75% of _____.

5.   7 is _____% of 16.   _____ is 25% of 64.

6.   30 is 60% of _____.   57 is _____% of 60.

7.   _____ is 75% of 60.   72 is 80% of _____.

8.   16 is _____% of 80.   _____ is $8\frac{1}{4}$% of 40.

9.   74 is 5% of _____.   85 is _____% of 170.

10.   _____ is $6\frac{1}{2}$% of 980.   968 is 100% of _____.

11.   117 is _____% of 78.   _____ is 150% of 90.

12.   42 is 30% of _____.   24 is _____% of 64.

Perfect score: 24   My score: _____

Problem Solving

Solve each problem.

1. An inspector found that 6% of the parts were faulty. One day 650 parts were made. How many of those parts are faulty?

_____ parts are faulty.

2. An order is for 1,000 parts. After 350 parts are made, what percent of the order has been made?

_____% of the parts has been made.

3. Mrs. Cook was able to purchase parts for her machine for 75% of the regular price. She paid $27 for the parts. What was the regular price of the parts?

The regular price was $_____.

4. There are 48 parking spaces in a parking lot. The lot is 62.5% filled. How many spaces are filled?

_____ spaces are filled.

5. When 42 out of the total number of parking spaces in problem 4 are filled, what percent of the spaces are filled?

_____% of the spaces are filled.

6. When the truck is loaded to 45% of capacity, there are 108 cases on the truck. How many cases will be on the truck when it is loaded to capacity?

_____ cases will be on the truck.

7. Jackie had 75 papers to sell. He has sold 45. What percent of the total number of papers has he sold?

He has sold _____% of the papers.

1.	
2.	**3.**
4.	**5.**
6.	**7.**

Perfect score: 7   My score: _____

106

CHAPTER 6 TEST

Complete the following.

a	*b*

1. _____ is 32% of 250.

   25 is _____% of 50.

2. _____ is 25% of 72.

   28 is 40% of _____.

3. 330 is 66% of _____.

   1.2 is _____% of 1.6.

4. 9 is _____% of 30.

   _____ is $6\frac{1}{2}$% of 900.

5. .8 is 40% of _____.

   _____ is 70% of 480.

6. 45 is _____% of 80.

   195 is 65% of _____.

7. _____ is $7\frac{1}{4}$% of 368.

   1.9 is _____% of 7.6.

8. 8 is 40% of _____.

   _____ is 150% of 90.

9. 7 is _____% of 7.

   1.4 is 70% of _____.

10. _____ is 9.2% of 680.

    210 is _____% of 240.

Perfect score: 20   My score: _____

107

PRE-TEST—Interest

Assume each principal has been loaned as indicated. Find the interest for each loan. Then find the total amount needed to repay each loan.

	principal	rate	time	interest	total amount
1.	$400	8%	1 year		
2.	$650	12%	1 year		
3.	$120	15%	1 year		
4.	$800	11%	2 years		
5.	$450	9%	3 years		

Complete the following.

	principal	rate	time	interest
6.	$250	10%	$1\frac{1}{2}$ years	
7.	$300	12%	$2\frac{1}{4}$ years	
8.	$180	9%	90 days	
9.	$60	8%	60 days	
10.	$200	14%	180 days	
11.	$180	13%	$1\frac{1}{2}$ years	
12.	$300	9%	$\frac{1}{2}$ year	
13.	$1000	$8\frac{1}{4}$%	$1\frac{1}{2}$ years	
14.	$1850	11%	3 years	
15.	$424	15%	$2\frac{3}{4}$ years	

Perfect score: 20   My score: _____

Lesson 1 Interest

Interest is the money paid for the use of money.

The amount of interest is determined by:

(1) the **principal,** the amount of money borrowed or deposited,

(2) the **rate** of interest, usually given as a percent, and

(3) the **time,** expressed in years.

When deposited in a savings account, what is the interest on $300 at 5% for 1 year?

$$interest = principal \times rate \times time$$

$$i = 300 \times .05 \times 1$$
$$= 15.00 \times 1$$
$$= 15.00 \text{ or } 15$$

The interest is $_____.

Complete the following.

	principal	rate	time	interest
1.	$200	5%	1 year	
2.	$300	4%	1 year	
3.	$400	5%	1 year	
4.	$75	6%	1 year	
5.	$30	7%	1 year	
6.	$78	4%	1 year	
7.	$830	4%	1 year	
8.	$925	6%	1 year	
9.	$42.80	5%	1 year	
10.	$513.40	5%	1 year	
11.	$64.50	8%	1 year	
12.	$1320.20	5%	1 year	

Perfect score: 12   My score: _____

Problem Solving

Solve each problem.

1. Mr. Murray borrowed $450 for 1 year. He is to pay interest at the rate of 9% yearly. How much interest will he pay?

He will pay $_____ interest.

2. Margaret put $65 in a savings account. The account will earn 6% interest yearly. How much interest will the account earn in 1 year?

It will earn $_____ interest.

3. Bob has a $25 bond that pays interest at the rate of 5% a year. How much interest will the bond pay after 1 year?

The bond will pay $_____ interest.

4. The Merkels borrowed $675 for 1 year. The interest rate is 14% a year. How much interest will the Merkels pay?

They will pay $_____ interest.

5. Mrs. Trumpet had $2,400 in her savings account at the beginning of the year. Her account pays 5% interest yearly. How much interest will she receive in 1 year?

She will receive $_____ in 1 year.

6. Assume the interest rate is 12% a year. How much interest would a person have to pay if he borrowed $420.75 for 1 year?

He would have to pay $_____ interest.

7. Al has $450 in an account that pays interest at 6% a year. Marge has $350 in an account that pays interest at 8% a year. Who will receive the greater amount of interest in 1 year? How much greater?

_____ will receive the greater amount.

It will be $_____ greater.

1.

2.	3.

4.	5.

6.	7.

Perfect score: 8   My score: _____

Lesson 2   Interest

You borrow $2,000 for a period of 3 years. You pay interest at the rate of $9\frac{3}{4}\%$ each year.

How much interest will you pay in the 3-year period?	How much is needed to repay the loan?
$i = p \times r \times t$ $= 2000 \times .0975 \times 3$ $= 585$	principal $2000 interest $+585$ total amount $2585
You will pay $ __585__ interest.	You need $ __2585__ to repay the loan.

Complete the following.

	principal	rate	time	interest
1.	$300	10%	3 years	
2.	$700	8%	2 years	
3.	$950	$12\frac{1}{2}\%$	4 years	
4.	$1480	$9\frac{1}{4}\%$	2 years	
5.	$1675	15%	3 years	

	principal	rate	time	interest	total amount
6.	$500	12%	2 years		
7.	$600	14%	3 years		
8.	$850	$9\frac{1}{2}\%$	2 years		
9.	$2500	$12\frac{3}{4}\%$	4 years		
10.	$3890	10%	3 years		

Perfect score: 15   My score: _____

Problem Solving

Solve each problem.

1. Juan borrowed $750 for 2 years. The interest rate is 9% a year. How much interest will he pay?

He will pay $_____ interest.

2. The principal of a loan is $2,500. The yearly interest rate is 8%. How much interest would be paid if the loan were for 3 years? How much would be needed to repay the loan?

The amount of interest is $_____.

The amount to repay is $_____.

3. Betty has a loan in the amount of $500. The interest rate is 10% a year. The loan is for 2 years. How much interest will she pay?

She will pay $_____ interest.

4. The Glazeskis borrowed $3,000 for 2 years. The yearly interest rate is 9%. How much interest will they pay?

They will pay $_____ interest.

5. Mrs. Henry has $5,500 in a savings account that pays $8\frac{1}{4}$% yearly interest. If the interest is mailed to her at the end of each year, how much will she receive in 3 years?

She will receive $_____ interest.

6. Suppose the interest rate in problem 5 is $8\frac{1}{2}$%. How much interest would Mrs. Henry receive in 3 years?

She would receive $_____ interest.

7. Millie borrowed $4,050 for 3 years at a yearly interest rate of 11%. How much will be needed to repay the loan?

The amount to repay is $_____.

1.	
2.	**3.**
4.	**5.**
6.	**7.**

Perfect score: 8   My score: _____

112

Lesson 3   Interest

You borrow $1,500 for a period of $2\frac{1}{2}$ years at a rate of 15% each year.

How much interest will you pay in the $2\frac{1}{2}$-year period?	How much is needed to repay the loan?
$i = p \times r \times t$ $\quad = 1500 \times .15 \times 2.5$ $\quad = 562.5$	principal $1500 interest $+562.50$ total amount $2062.50
You will pay $ _562.50_ interest.	You need $ _2062.50_ to repay the loan.

Complete the following.

	principal	rate	time	interest
1.	$500	12%	2 years	
2.	$450	8%	$2\frac{1}{2}$ years	
3.	$392	15%	$1\frac{1}{4}$ years	
4.	$1870	$10\frac{1}{2}$%	3 years	
5.	$2000	$11\frac{1}{4}$%	$2\frac{3}{4}$ years	

	principal	rate	time	interest	total amount
6.	$700	10%	$1\frac{1}{2}$ years		
7.	$85	12%	$\frac{1}{2}$ year		
8.	$1000	$9\frac{1}{2}$%	$\frac{3}{4}$ year		
9.	$1875	14%	$3\frac{1}{2}$ years		
10.	$1400	$9\frac{3}{4}$%	$2\frac{1}{2}$ years		

Perfect score: 15   My score: _____

Problem Solving

Solve each problem.

1. Mr. Rogers borrowed $800 for a period of 2 years. He is to pay interest at the rate of 9% a year. How much interest will he have to pay?

He will have to pay $_____ interest.

2. Roger had $445 in a savings account that paid interest at the rate of 6% a year. How much interest did Roger receive in $1\frac{1}{2}$ years?

Roger received $_____ interest.

3. Mrs. Lazar deposited $160 in an account that pays $6\frac{1}{2}$% interest each year. How much interest would she have after $2\frac{1}{2}$ years?

She would have $_____ interest.

4. Mr. Wrinkles borrowed $750 for $\frac{1}{2}$ year. How much interest will he be charged if the interest rate is 8% a year? What is the total amount needed to repay the loan?

He will be charged $_____ interest.

The total amount needed is $_____.

5. The Mims borrowed $3000 for $2\frac{1}{2}$ years at $12\frac{3}{4}$% yearly interest. What is the total amount needed to repay the loan?

The total amount needed is $_____.

6. Marvin has $360 in an account that pays 5% interest a year. Bill has $290 in an account that pays 6% interest a year. Who will receive the greater amount of interest after $1\frac{1}{2}$ years? How much greater?

_____ will receive the greater amount.

He will receive $_____ more.

1.	2.
3.	**4.**
5.	**6.**

Perfect score: 8   My score: _____

114

Lesson 4   Interest

When computing interest for a certain number of days, a year is usually considered to be 360 days. Thus, 90 days is $\frac{90}{360}$ or $\frac{1}{4}$ year, 180 days is $\frac{180}{360}$ or $\frac{1}{2}$ year, and so on.

How much interest would Mrs. Willis pay for a 30-day loan of $600 at 8%?

$$i = 600 \times .08 \times \tfrac{30}{360}$$
$$= 48 \times \tfrac{1}{12}$$
$$= 4$$

She would pay $_____4_____ interest.

How much interest would Mrs. Willis pay for a 120-day loan of $600 at 8%?

$$i = 600 \times .08 \times \tfrac{120}{360}$$
$$= 48 \times \tfrac{1}{3}$$
$$= 16$$

She would pay $_____ interest.

Find the interest for each of the following.

	principal	rate	time	interest
1.	$200	12%	30 days	
2.	$240	10%	60 days	
3.	$150	8%	90 days	
4.	$300	15%	120 days	
5.	$420	11%	180 days	
6.	$900	13%	270 days	
7.	$800	15%	30 days	
8.	$900	12%	180 days	
9.	$540	10%	80 days	
10.	$1900	15%	150 days	

Perfect score: 10   My score: _____

Problem Solving

Solve each problem.

1. Interest on a 90-day loan of $400 was charged at the rate of 15% a year. How much interest was charged?

The interest would be $_____.

2. Suppose $350 is invested for 60 days at an annual (yearly) rate of 6%. How much interest will be earned?

$_____ interest would be earned.

3. The Triangle Company borrowed $1900 for 180 days at 12% annual interest. How much interest will they have to pay? What will be the total amount needed to repay the loan?

The interest will be $_____.

The total amount will be $_____.

4. Mr. Davis borrowed $600 for 60 days at 9% annual interest. However, he was able to repay the loan in 30 days. How much interest was he able to save by doing this?

He was able to save $_____ interest.

5. Interest on a 120-day loan of $27,000 is charged at an annual rate of 10%. How much interest is charged?

The interest is $_____.

6. An automobile dealer borrowed $36,000 from the bank at 9% annual interest. How much interest will be charged for 270 days? If the loan is paid in 270, what will be the total amount needed to repay the loan?

He will be charged $_____ interest.

The total amount will be $_____.

1.	2.
3.	**4.**
5.	**6.**

Perfect score: 8   My score: _____

CHAPTER 7 TEST

Assume each principal has been loaned as indicated. Find the interest for each loan. Then find the total amount needed to repay each loan.

	principal	rate	time	interest	total amount
1.	$500	9%	1 year		
2.	$400	11%	1 year		
3.	$600	15%	2 years		
4.	$750	8%	3 years		
5.	$3000	$9\frac{1}{2}$%	2 years		

Complete the following.

	principal	rate	time	interest
6.	$300	15%	$1\frac{1}{2}$ years	
7.	$500	11%	$2\frac{1}{4}$ years	
8.	$400	12%	$2\frac{3}{4}$ years	
9.	$650	14%	180 days	
10.	$700	$8\frac{1}{4}$%	2 years	
11.	$800	10%	90 days	
12.	$1000	$11\frac{1}{2}$%	$\frac{1}{2}$ year	
13.	$880	$7\frac{3}{4}$%	2 years	
14.	$1550	16%	270 days	
15.	$1600	14%	4 years	

Perfect score: 20   My score: _____

PRE-TEST—Metric Measurement

Complete the following.

a   *b*

1. 9 centimeters = _____ millimeters   1 meter = _____ millimeters

2. 25 meters = _____ centimeters   1 kilometer = _____ meters

3. 30 millimeters = _____ centimeters   3500 millimeters = _____ meters

4. 2.5 kilometers = _____ meters   46.87 meters = _____ millimeters

5. 49 centimeters = _____ meter   250 meters = _____ kilometer

6. 4 liters = _____ milliliters   1.7 kiloliters = _____ liters

7. 500 milliliters = _____ liter   2480 liters = _____ kiloliters

8. 10.8 grams = _____ milligrams   800 milligrams = _____ grams

9. 100 kilograms = _____ grams   260.5 grams = _____ kilogram

10. 4800 kilograms = _____ metric tons   .9 metric ton = _____ kilograms

Solve each problem.

11. Robin needs 5 meters of ribbon. The ribbon comes in rolls of 250 centimeters each. How many rolls will she have to buy?

She will have to buy _____ rolls.

12. 12 servings of the same size were made from a box of cereal that weighs 336 grams. How much cereal was in each serving?

There were _____ grams in each serving.

13. Mrs. Cardenal's car can go 11.3 kilometers on 1 liter of fuel. The capacity of the car's fuel tank is 80 liters. How far can she drive on a full tank of fuel?

She can drive _____ kilometers.

11.	
12.	13.

Perfect score: 23   My score: _____

Lesson 1   Length

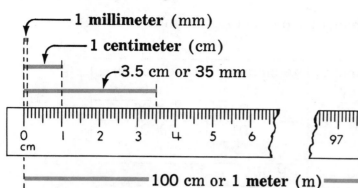

1cm	= 10 mm	1 mm	= .1 cm
1 m	= 100 cm	1 mm	= .001 m
1 m	= 1000 mm	1 cm	= .01 m
1 km	= 1000 m	1 m	= .001 km

A distance of 1,000 meters is 1 kilometer (km). If you run around a football field (including end zones) 3 times, you would run a distance of about 1 kilometer.

16.9 m = ___?___ cm	.8 m = ___?___ mm
16.9 m = (16.9 × 100) cm	.8 m = (.8 × 1000) mm
16.9 m = __1690__ cm	.8 m = _____ mm

Complete the following.

	a	*b*	*c*
1.	9 km = _____ m	12 m = _____ cm	6 m = _____ mm
2.	3 cm = _____ mm	5.3 km = _____ m	.4 m = _____ cm
3.	9.2 cm = _____ mm	.99 m = _____ mm	10 km = _____ m
4.	10.54 km = _____ m	12.8 m = _____ cm	1.2 m = _____ mm
5.	.01 m = _____ cm	100 m = _____ mm	45 cm = _____ mm
6.	200 km = _____ m	18.7 cm = _____ mm	.01 km = _____ m

Solve each problem.

7. The diameter of a hockey puck is 76 millimeters. Is that more or is that less than 8 centimeters?

It is _____ than 8 centimeters.

8. A relay race is 1.5 kilometers long. How many meters is that?

The race is _____ meters long.

7.

8.

Perfect score: 20   My score: _____

Lesson 2 Units of Length

This table shows how to change a measurement from one metric unit to another.

To change	to millimeters, multiply by	to centimeters, multiply by	to meters, multiply by	to kilometers, multiply by
millimeters		.1	.001	.000001
centimeters	10		.01	.00001
meters	1000	100		.001
kilometers	1,000,000	100,000	1000	

5300 mm = ___?___ cm
5300 mm = (5300 × .1) cm
5300 mm = __530.0__ cm

420.8 m = ___?___ km
420.8 m = (420.8 × .001) km
420.8 m = __.4208__ km

Complete the following.

a   b   c

1.   130 mm = _____ cm   2.8 cm = _____ m   345 m = _____ km

2.   60 cm = _____ mm   100 mm = _____ m   9.25 km = _____ m

3.   .9 cm = _____ m   2.05 km = _____ m   1500 m = _____ km

4.   20.6 cm = _____ m   24.85 m = _____ mm   48 mm = _____ cm

5.   999 mm = _____ m   98.9 m = _____ cm   .1 cm = _____ mm

6.   50 m = _____ km   152.6 cm = _____ m   .625 km = _____ m

Solve each problem.

7. The circumference of a baseball is 238 millimeters. How many centimeters is that?

The circumference is _____ centimeters.

8. Adlai said he is 150 millimeters tall. Alex said he is 150 centimeters tall. Kathy said she is 150 meters tall. Only one of those pupils could possibly be correct. What is the name of that pupil?

_____ gave a correct measurement.

7.

8.

Perfect score: 20   My score: _____

Lesson 3   Capacity

NAME _____

A cube with each edge 1 centimeter long has a capacity of 1 **milliliter** (ml). A cube that is 10 centimeters long on each edge has a capacity of 1 **liter**.

1 liter = 1000 ml
1 ml   = .001 liter

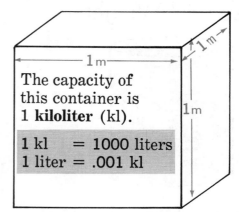

The capacity of this container is 1 **kiloliter** (kl).

1 kl   = 1000 liters
1 liter = .001 kl

To change	to milliliters, multiply by	to liters, multiply by	to kiloliters, multiply by
milliliters		.001	.000001
liters	1000		.001
kiloliters	1,000,000	1000	

Complete the following.

a   *b*

1. 8 kl = _____ liters   17 kl = _____ liters

2. 200 liters = _____ kl   6 liters = _____ ml

3. 2000 ml = _____ liters   3440 liters = _____ kl

4. 10 liters = _____ ml   45 ml = _____ liter

5. .1 kl = _____ liters   .94 liter = _____ ml

6. 240.5 liters = _____ kl   236 ml = _____ liter

7. 3500 ml = _____ liter   4.5 kl = _____ liters

8. 1 liter = _____ kl   30.8 liters = _____ ml

Solve each problem.

9. There are 1,000 cubic centimeters in 1 liter. How many cubic centimeters are there in 1 kiloliter?

_____ cubic centimeters are in 1 kiloliter.

10. An owner of a service station bought 3 kiloliters of gasoline at $.34 a liter. How much did it cost?

The gasoline cost $_____.

9.

10.

Perfect score: 18   My score: _____

121

Lesson 4   Weight

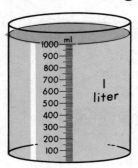

1 liter of water
weighs 1 **kilogram** (kg).

1 milliliter of water
weighs 1 **gram** (g).

1 drop of water weighs
about 67 **milligrams** (mg).

| 1 kg = 1000 g | 1 g = .001 kg |
| 1 g = 1000 mg | 1 mg = .001 g |

Complete the following.

	a	b
1.	3 kg = _____ g	10 g = _____ mg
2.	485 g = _____ kg	250 mg = _____ g
3.	3.5 kg = _____ g	4600 mg = _____ g
4.	45.8 g = _____ mg	2500.9 g = _____ kg

Solve each problem.

5. Give the weight in kilograms of 2.5 liters of water.
Give the weight in grams.

The weight is _____ kilograms.

The weight is _____ grams.

6. Give the weight in grams of 9.5 milliliters of water. Give the weight in milligrams.

The weight is _____ grams.

The weight is _____ milligrams.

7. A vitamin tablet weighs 100 milligrams. How many vitamin tablets would it take to weigh 1 gram?

It would take _____ vitamin tablets.

5.

6.   **7.**

Perfect score: 13   My score: _____

122

Lesson 5 Metric Ton

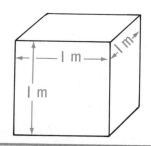

Suppose this container were filled with water. The water would weigh 1 **metric ton** (t).

1 t = 1000 kg	1 kg = .001 t

Complete the following.

          a   *b*

1. 6 t = _____ kg   8.5 t = _____ kg

2. 675 kg = _____ t   5050.5 kg = _____ t

Solve each problem.

3. Peggy weighs 49.3 kilograms. Amy weighs 47.7 kilograms. How much more does Peggy weigh than Amy?

Peggy weighs _____ kilograms more.

3.

4. A bag of apples weighing 2.6 kilograms costs $2.99. What is the cost per kilogram?

The cost per kilogram is $_____.

5. A kiloliter of water weighs 1 metric ton. There are 7,500 kiloliters of water in a swimming pool. How much does the water in the swimming pool weigh?

The water weighs _____ metric tons.

4.   **5.**

6. The combined weight of the pupils in a class is 1,260 kilograms. Give the weight in metric tons.

The weight is _____ metric tons.

6.   **7.**

7. The United States produces about 150,000 metric tons of wheat a day. It produces about 200,000 metric tons more of corn a day than wheat. How many metric tons of corn does the United States produce a day?

It produces _____ metric tons of corn a day.

Perfect score: 9   My score: _____

Problem Solving

Solve each problem.

1. A roll has 325 centimeters of tape. How many meters is that? How many millimeters is that?

It is _____ meters.

It is _____ millimeters.

2. Bonnie Mae runs 3.5 kilometers each day. How many meters does she run each day?

She runs _____ meters each day.

3. A teaspoon holds 5 milliliters of water. How many teaspoonfuls of water are in 1 liter?

_____ teaspoonfuls of water are in 1 liter.

4. 1.5 liters of water are in a container. Suppose 50 milliliters of water are removed. How many liters of water would remain in the container?

_____ liters of water would remain.

5. It costs $1.86 for 2 liters of orange juice. What is the cost per liter?

The cost per liter is $_____.

6. It costs $1.80 to buy 2.25 kilograms of onions. What is the cost per kilogram?

The cost per kilogram is $_____.

7. 1 nickel weighs about 5 grams. There are 40 nickels in a roll. Find the weight in grams of 3 rolls of nickels.

The weight is _____ grams.

8. There are 3.5 metric tons of cargo on a truck. How many kilograms is that?

It is _____ kilograms.

1.	2.
3.	**4.**
5.	**6.**
7.	**8.**

Perfect score: 9   My score: _____

124

CHAPTER 8 TEST

Complete the following.

	a	*b*
1.	12 cm = _____ mm	6 m = _____ mm
2.	30 m = _____ cm	2 km = _____ m
3.	80 mm = _____ cm	900 mm = _____ m
4.	6.4 km = _____ m	54.92 m = _____ mm
5.	54.92 cm = _____ m	505 m = _____ km
6.	6 liters = _____ ml	1.2 kl = _____ liters
7.	.7 liter = _____ ml	85 ml = _____ liter
8.	1200 ml = _____ liters	3500 liters = _____ kl
9.	15.9 g = _____ mg	75 mg = _____ g
10.	250 kg = _____ g	875 g = _____ kg
11.	10,000 kg = _____ t	.75 t = _____ kg

Solve each problem.

12. The odometer on a car showed 9031.7 kilometers before a trip. It showed 10,001.3 kilometers after the trip. How long was the trip?

The trip was _____ kilometers long.

13. 1 liter of water weighs 1 kilogram. How many grams do 1.5 liters of water weigh?

The water weighs _____ grams.

14. A pharmacist has 700 milliliters of medicine. She wants to put the same amount of medicine in each of 35 bottles. How much medicine will she put in each bottle?

She will put _____ milliliters in each bottle.

12.	
13.	14.

Perfect score: 25   My score: _____

NAME _____

PRE-TEST—Geometry

Name each figure below as shown.

    a   *b*   *c*   *d*

1.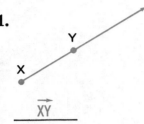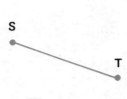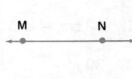

$\overrightarrow{XY}$ _____ $\angle PQR$ _____ $\overline{ST}$ _____ $\overleftrightarrow{MN}$ _____

2. 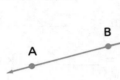 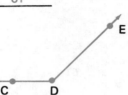 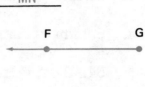 (images)

_____   _____   _____   _____

Use a ruler to compare the lengths of the sides of each triangle. Then tell whether it is a *scalene,* an *isosceles,* or an *equilateral* triangle.

    a   *b*   *c*

3.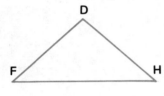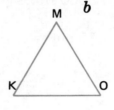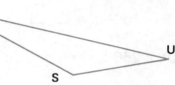

_____ triangle   _____ triangle   _____ triangle

Write an *R* in each rectangle below. Write an *S* in each square. Write an *X* in each rhombus.

4.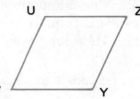

Tell whether the following are *parallel lines* or *intersecting lines*.

5.  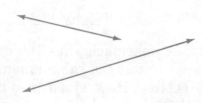 (images)

_____   _____   _____

Perfect score: 15   My score: _____

Lesson 1   Points, Lines, and Line Segments

P
•

A point can be represented by a dot. Point P is shown above.

A ━━━━━━━━━━ B

Line AB (denoted $\overleftrightarrow{AB}$) names the line which passes through points A and B. Does $\overleftrightarrow{BA}$ name the same line as $\overleftrightarrow{AB}$?_____

M ━ ━ ━━━━━━━ N

Line segment MN (denoted $\overline{MN}$) consists of points M and N and all points on the line between M and N. Does $\overline{NM}$ name the same segment as $\overline{MN}$? _____

Complete the following as shown.

		a		*b*	

1. ●━━━● line __CD__ or __DC__ $\overleftrightarrow{CD}$ or $\overleftrightarrow{DC}$
       C   D

2. S ╲ line ___ or ___   ___ or ___
        ╲ R

3. ╱ X line ___ or ___   ___ or ___
     Y

4. ●━━━● line ___ or ___   ___ or ___
    P   Q

Complete the following as shown.

		a		*b*		*c*

5. ●━━━● line segment __EF__ or __FE__ $\overline{EF}$ or $\overline{FE}$ endpoints: __E__ and __F__
    E   F

6. J ●╲ line segment ___ or ___   ___ or ___   endpoints: ___ and ___
        ╲ K

7. ●━━━● line segment ___ or ___   ___ or ___   endpoints: ___ and ___
    R   S

8. ●━━━● G line segment ___ or ___   ___ or ___   endpoints: ___ and ___
    B

Perfect score: 30   My score: _____

Lesson 2   Rays and Angles

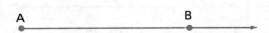

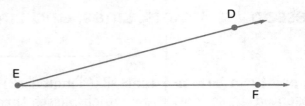

Ray AB (denoted $\overrightarrow{AB}$) consists of point A and all points on $\overleftrightarrow{AB}$ that are on the same side of A as B. The endpoint of $\overrightarrow{AB}$ is point _____.

An **angle** is formed by two rays which have a common endpoint. Angle DEF (denoted ∠DEF) is formed by rays $\overrightarrow{ED}$ and $\overrightarrow{EF}$. Does ∠FED name the same angle? _____

Complete the following as shown.

1. ●————————● →   ray __MN__   __$\overrightarrow{MN}$__   endpoint of ray: __M__
    M   N

2.   Q   ray _____   _____   endpoint of ray: _____
       P

3. X   ray _____   _____   endpoint of ray: _____
              Y

4.   ray _____   _____   endpoint of ray: _____
       J   K

5. A   angle ___ABC___ or ___CBA___   ___∠ABC___ or ___∠CBA___

   B   C   rays __$\overrightarrow{BA}$__ and __$\overrightarrow{BC}$__

6.   L   angle _____ or _____   _____ or _____

   J   K   rays _____ and _____

7.   Q   angle _____ or _____   _____ or _____
   P   R
                                 rays _____ and _____

Perfect score: 21   My score: _____

128

Lesson 3   Parallel and Intersecting Lines

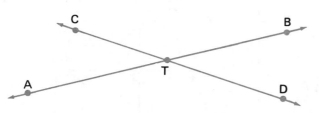

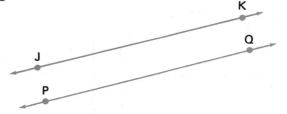

Lines like $\overleftrightarrow{AB}$ and $\overleftrightarrow{CD}$ are called **intersecting lines.** What point do $\overleftrightarrow{AB}$ and $\overleftrightarrow{CD}$ have in common? _____

Lines like $\overleftrightarrow{JK}$ and $\overleftrightarrow{PQ}$ are called **parallel lines.** Will $\overleftrightarrow{JK}$ and $\overleftrightarrow{PQ}$ ever intersect, no matter how far extended? _____

Complete the following as shown.

	a type of lines		*b* type of lines
1.	parallel		intersecting
2.	_____		_____
3.	_____		_____

Answer the following.

4. In how many points do two parallel lines intersect?   _____

5. Can two lines be parallel and also intersect?   _____

6. In how many points can two lines intersect?   _____

7. If $\overleftrightarrow{AB}$ is parallel to $\overleftrightarrow{CD}$, is $\overleftrightarrow{CD}$ parallel to $\overleftrightarrow{AB}$?   _____

Perfect score: 8   My score: _____

Lesson 4   Right Angles

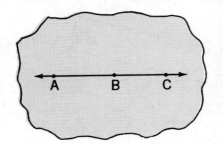

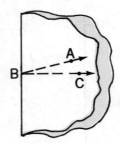

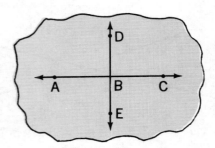

Draw a line on a sheet of paper. Mark points A, B, and C as shown above.

Fold the line over itself at B.

Unfold the paper. Draw a line through B along the fold line. Label it as shown. Line AC and line DE form four **right angles**.

Angles such as ∠DBA, ∠DBC, ∠ABE, and ∠CBE are right angles. Two lines that form right angles are called **perpendicular lines.**

Write an *R* beside each right angle.

| *a* | *b* | *c* | *d* |

1.

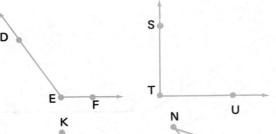

2.

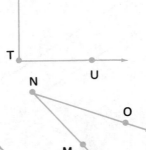

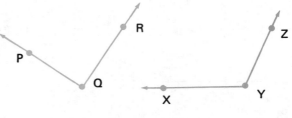

Write *P* beside each pair of perpendicular lines.

| *a* | *b* | *c* | *d* |

3.

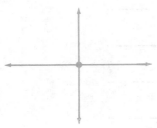

Perfect score: 12   My score: _____

130

Lesson 5   Types of Angles

Compare ∠ABC with a model of a right angle, such as the corner of a sheet of paper.

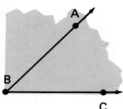

Does ∠ABC appear to be larger or smaller than

a right angle? _____ Angles like ∠ABC are called **acute angles.**

Compare ∠PQR with a model of a right angle.

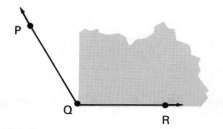

Does ∠PQR appear to be larger or smaller than

a right angle? _____ Angles like ∠PQR are called **obtuse angles.**

Compare each angle with a model of a right angle. Then tell whether the angle is an *acute*, an *obtuse*, or a *right* angle.

| | *a* | *b* | *c* |

1.

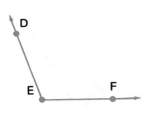

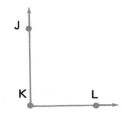

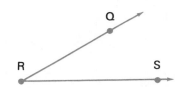

_____   _____   _____

2.

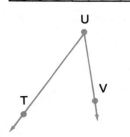

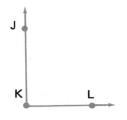

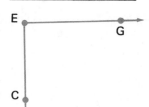

_____   _____   _____

3.

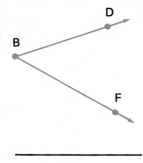

 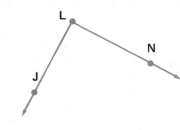

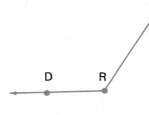

_____   _____   _____

Perfect score: 9   My score: _____

Lesson 6   Types of Triangles by Angles

Compare the angles of each triangle with a model of a right angle.

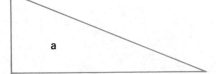
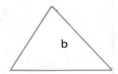
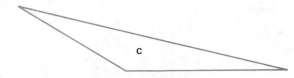

An **acute triangle** contains all acute angles.

   Which triangle above is an acute triangle? _____

A **right triangle** contains one right angle.

   Which triangle above is a right triangle? _____

An **obtuse triangle** contains one obtuse angle.

   Which triangle above is an obtuse triangle? _____

Compare the angles of each triangle with a model of a right angle. Then tell whether the triangle is an *acute,* an *obtuse,* or a *right* triangle.

a	b	c

1.

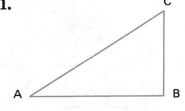
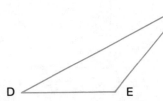
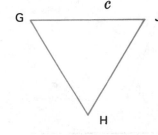

_____ triangle   _____ triangle   _____ triangle

2.

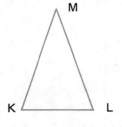
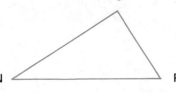
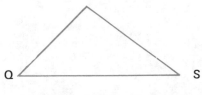

_____ triangle   _____ triangle   _____ triangle

3.

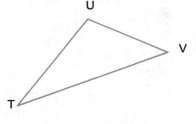
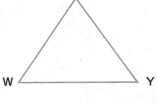
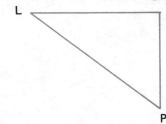

_____ triangle   _____ triangle   _____ triangle

Perfect score: 9   My score: _____

Lesson 7 Types of Triangles by Sides

Use a ruler to compare the lengths of the sides of each triangle.

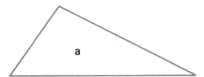

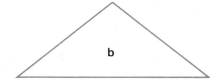

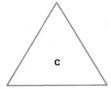

In a **scalene triangle** no two sides are congruent.

Which triangle above is a scalene triangle? _____

In an **isosceles triangle** at least two sides are congruent.

Which triangles above are isosceles triangles? _____

In an **equilateral triangle** all sides are congruent.

Which triangle above is an equilateral triangle? _____

Use a ruler to compare the lengths of the sides of each triangle. Then tell whether the triangle is a *scalene*, an *isosceles*, or an *equilateral* triangle.

a	*b*	*c*

1.

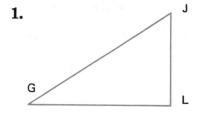

_____ triangle

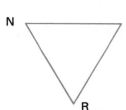

_____ triangle

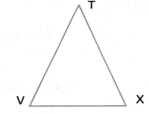

_____ triangle

2.

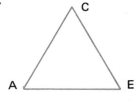

_____ triangle

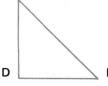

_____ triangle

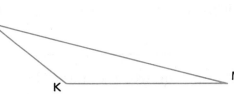

_____ triangle

3.

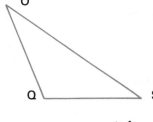

_____ triangle

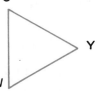

_____ triangle

_____ triangle

Perfect score: 9 / My score: _____

Lesson 8 Types of Quadrilaterals

A parallelogram is a quadrilateral (4-sided figure) in which the opposite sides are parallel.

$\overrightarrow{AB}$ is parallel to $\overrightarrow{DC}$ and $\overrightarrow{AD}$ is parallel to $\overrightarrow{BC}$.

Is figure ABCD a parallelogram? _____

A rectangle is a quadrilateral in which all angles are right angles.

Compare each of the angles of figure EFGH with a model of a right angle.

Is figure EFGH a rectangle? _____

A rhombus is a quadrilateral in which all sides are congruent.

Use a ruler to compare the lengths of the sides of figure JKLM.

Is figure JKLM a rhombus? _____

A square is a quadrilateral in which all angles are right angles and all sides are congruent.

Compare the angles of figure PQRS with a model of a right angle and use a ruler to compare the lengths of its sides.

Is figure PQRS a square? _____

Use the figures at the top of the page to answer the questions which follow.

a	b
1. Which figures are rectangles?	Which figure is a square?
_____	_____
2. Are all squares rectangles?	Are all rectangles squares?
_____	_____
3. Which figures are rhombuses?	Which figure is a square?
_____	_____
4. Are all squares rhombuses?	Are all rhombuses squares?
_____	_____

Perfect score: 8   My score: _____

134

CHAPTER 9 TEST

Name each figure.

a   b   c

1.    

_____   _____   _____

Compare the angles of each triangle with a model of a right angle. Then tell whether the triangle is an *acute,* an *obtuse,* or a *right* triangle.

a   b   c

2. 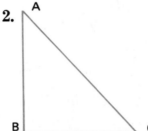   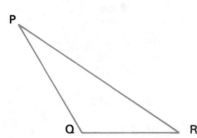   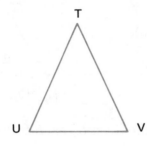

_____ triangle   _____ triangle   _____ triangle

In each figure below, the opposite sides are parallel. Use these figures to answer the questions that follow.

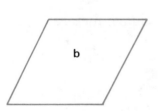 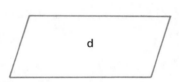

3. Which figures are squares? _____

4. Which figures are rectangles? _____

5. Which figures are parallelograms? _____

6. Which figures are rhombuses? _____

Perfect score: 10   My score: _____

PRE-TEST—Perimeter and Area

Find the perimeter (or circumference) of each figure below. Use $3\frac{1}{7}$ for π.

a	*b*	*c*

1.

3 cm

12 cm

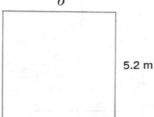

5.2 m

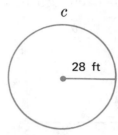

28 ft

_____ cm   _____ m   _____ ft

2.

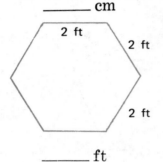

2 ft   2 ft   2 ft

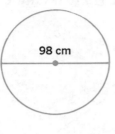

98 cm

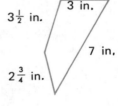

3 in.

$3\frac{1}{2}$ in.

7 in.

$2\frac{3}{4}$ in.

_____ ft   _____ cm   _____ in.

Find the area of each figure below. Use 3.14 for π.

3.

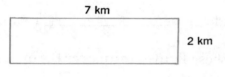

7 km

2 km

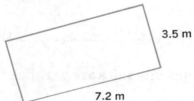

3.5 m

7.2 m

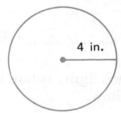

4 in.

_____ square kilometers   _____ square meters   _____ square inches

4.

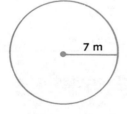

7 m

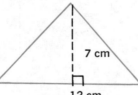

7 cm

12 cm

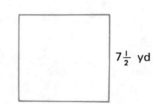

$7\frac{1}{2}$ yd

_____ square meters   _____ square centimeters   _____ square yards

5.

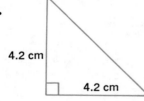

4.2 cm

4.2 cm

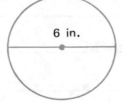

6 in.

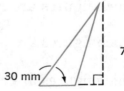

70 mm

30 mm

_____ square centimeters   _____ square inches   _____ square millimeters

Perfect score: 15   My score: _____

Lesson 1   Perimeter

The distance around a figure is called its perimeter.

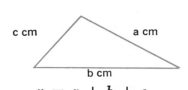

$p = a + b + c$

If $a = 4$, $b = 5$, and $c = 3$, the perimeter of the tri-

angle is _____ centimeters.

$p = 4 \times s$

If $s = 7\frac{1}{2}$, the perimeter

of the square is _____ yd.

$p = 2 \times (l + w)$

If $l = 8$ and $w = 3$, the perimeter of the **rectangle**

is _____ meters.

Find the perimeter of each figure.

a	*b*	*c*

1.

7 cm

3 cm

_____ cm

2 ft

5 ft

5 ft

_____ ft

3 m

_____ m

2.

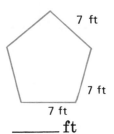

7 ft

7 ft

7 ft

_____ ft

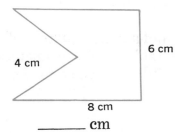

4 cm

6 cm

8 cm

_____ cm

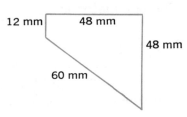

12 mm   48 mm

48 mm

60 mm

_____ mm

Complete the table for each rectangle described below.

	length	width	perimeter
3.	9 m	4 m	_____ m
4.	$4\frac{1}{2}$ ft	$3\frac{1}{2}$ ft	_____ ft
5.	7.5 cm	2.5 cm	_____ cm
6.	$3\frac{1}{2}$ mi	2 mi	_____ mi
7.	7.5 ft	6.5 ft	_____ ft

Perfect score: 11   My score: _____

Problem Solving

Solve each problem.

1. The two longer sides of a lot measure 204 feet and 192 feet. The two shorter sides measure 92 feet and 54 feet. What is the perimeter of the lot?

The perimeter is _____ feet.

2. Mrs. Batka's lot is shaped like a rectangle. It is 175 feet long and 48 feet wide. What is the perimeter of her lot?

The perimeter is _____ feet.

3. A photograph measures $3\frac{1}{2}$ inches on each side. What is its perimeter?

Its perimeter is _____ inches.

4. A flower bed is in the shape of a triangle. The sides measure $2\frac{1}{2}$ yards, $3\frac{1}{2}$ yards, and $4\frac{1}{4}$ yards. What is the perimeter of the flower bed?

The perimeter is _____ yards.

5. Al is going to put tape around a rectangular table. He has 2.5 meters of tape. The table is 60 centimeters wide and 70 centimeters long. Will the tape be too long or too short? By how much?

The tape will be too _____

by _____ centimeters.

6. Each side of a triangle is 2.3 meters long. Each side of a square is 1.6 meters long. Which figure has the greater perimeter? How much greater?

The _____ has the greater perimeter.

It is _____ meter greater.

Perfect score: 8   My score: _____

138

Lesson 2 Circumference of a Circle

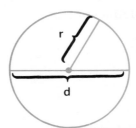

The perimeter of a circle is called its **circumference**.
The length of a *diameter* is twice the length of a *radius*.

$$d = 2 \times r \text{ and } r = d \div 2$$

For circles of all sizes, $C \div d$ names the same number. This number is represented by π (spelled *pi* and pronounced *pie*). π is approximately equal to $3\frac{1}{7}$ or 3.14.

To find the circumference of a circle, use $C = \pi \times d$ or $C = 2 \times \pi \times r$.

Find the circumference of each circle. Use $3\frac{1}{7}$ for π.

	a	*b*	*c*

1.

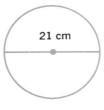   21 cm

   7 m

10 yd

_____ cm   _____ m   _____ yd

2.

 $5\frac{1}{4}$ in.

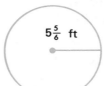 $5\frac{5}{6}$ ft

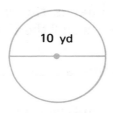 $3\frac{1}{2}$ in.

_____ in.   _____ ft   _____ in.

Complete the table for each circle described below. Use 3.14 for π.

	diameter	*radius*	*circumference*
3.	8 m	_____ m	_____ m
4.	_____ ft	3 ft	_____ ft
5.	7 in.	_____ in.	_____ in.
6.	_____ cm	8.5 cm	_____ cm
7.	9.4 km	_____ km	_____ km
8.	_____ m	7.2 m	_____ m

Perfect score: 18   My score: _____

Problem Solving

Solve each problem. Use $3\frac{1}{7}$ for π in problems **1-3**. Use 3.14 for π in problems **4-6**.

1. The wheel on Mr. Wilson's car has a 7-inch radius. What is the diameter of the wheel? What is the circumference of the wheel?

The diameter is _____ inches.

The circumference is _____ inches.

2. The tire on Miss Wolfson's car has a 28-inch diameter. What is the radius of the tire? What is the circumference of the tire?

The radius is _____ inches.

The circumference is _____ inches.

3. A circular play area has a radius of $16\frac{1}{2}$ yards. What is the circumference of the play area?

The circumference is _____ yards.

4. The diameter of a dime is 18 millimeters. How long is the radius? How long is the circumference?

The radius is _____ millimeters.

The circumference is _____ millimeters.

5. The lid of a can has a 2.5-meter diameter. What is the circumference of the lid?

The circumference is _____ meters.

6. Which has the greater perimeter, the square or the circle? How much greater?

3 m

The _____ has the greater perimeter.

It is _____ meters greater.

1.	
2.	
3.	**4.**
5.	**6.**

Perfect score: 10   My score: _____

Lesson 3   Circumference of a Circle

$C = \pi \times d$ and $d = C \div \pi$	$C = 2 \times \pi \times r$ and $r = C \div (2 \times \pi)$
Find d if $C = 66$.	Find r if $C = 25.12$.

Find d if $C = 66$.

$$d = C \div \pi$$
$$= 66 \div \tfrac{22}{7}$$
$$= 66 \times \tfrac{7}{22}$$
$$= 21$$

The diameter is _____ units.

Find r if $C = 25.12$.

$$r = C \div (2 \times \pi)$$
$$= 25.12 \div (2 \times 3.14)$$
$$= 25.12 \div 6.28$$
$$= 4$$

The radius is _____ units.

Complete the table for each circle described below. Use $3\tfrac{1}{7}$ for π in problems **1-6**. Use 3.14 for π in problems **7-12**.

	diameter	radius	circumference
1.	_____ in.	_____ in.	22 in.
2.	_____ ft	_____ ft	110 ft
3.	_____ in.	_____ in.	88 in.
4.	_____ yd	_____ yd	11 yd
5.	_____ ft	_____ ft	2 ft
6.	_____ yd	_____ yd	33 yd
7.	_____ m	_____ m	6.28 m
8.	_____ mm	_____ mm	31.4 mm
9.	_____ cm	_____ cm	9.42 cm
10.	_____ km	_____ km	21.98 km
11.	_____ m	_____ m	3.14 m
12.	_____ km	_____ km	12.56 km

Perfect score: 24   My score: _____

Problem Solving

Solve each problem. Use $3\frac{1}{7}$ for π in problems **1-3**. Use 3.14 for π in problems **4-6**.

1. The circumference of a wheel is 110 inches. What is the diameter of the wheel? What is the radius?

The diameter is _____ inches.

The radius is _____ inches.

2. What is the radius of a circle if the circumference is 7 feet? What is the diameter?

The radius is _____ feet.

The diameter is _____ feet.

3. The circumference of a coin is $3\frac{1}{7}$ inches. What is the diameter of the coin? What is the radius?

The diameter is _____ inch.

The radius is _____ inch.

4. The circumference of a circular flower bed is 9.42 meters. What is the radius of the flower bed? What is the diameter of the flower bed?

The radius is _____ meters.

The diameter is _____ meters.

5. The circumference of a circle is 1.57 centimeters. What is the diameter? What is the radius?

The diameter is _____ centimeter.

The radius is _____ centimeter.

6. The circumference of the circle is 9.72 meters less than the perimeter of the square. What is the diameter of the circle? What is the radius?

← 4 m →

The diameter is _____ meters.

The radius is _____ meter.

1.	
2.	
3.	
4.	
5.	
6.	

Perfect score: 12   My score: _____

142

Lesson 4   Area of Rectangles

To find the **area measure** (A) of a rectangle, multiply the measure of the *length* (l) by the measure of the *width* (w).

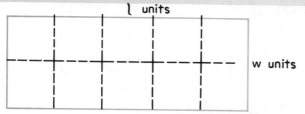

Find A if $l=5$ and $w=2$.
$$A=l\times w$$
$$=5\times \underline{\quad 2 \quad}$$
$$=\underline{\quad 10 \quad}$$

The area is __10__ square units.

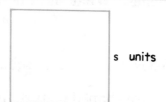

Find A if $s=7$.
$$A=s\times s$$
$$=\underline{\qquad}\times\underline{\qquad}$$
$$=\underline{\qquad}$$

The area is _____ square units.

Find the area of each rectangle below.

	a	*b*	*c*

1.

 8 ft / 4 ft

 8 in.

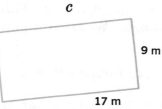 9 m / 17 m

_____ square feet   _____ square inches   _____ square meters

Complete the table for each rectangle described below.

	length	*width*	*area*
2.	18 mm	3 mm	_____ square millimeters
3.	7.5 cm	7.5 cm	_____ square centimeters
4.	$4\frac{1}{2}$ in.	$3\frac{1}{2}$ in.	_____ square inches
5.	$2\frac{1}{4}$ yd	$2\frac{1}{4}$ yd	_____ square yards
6.	6.5 m	2.5 m	_____ square meters
7.	8.3 km	4.1 km	_____ square kilometers

Perfect score: 9   My score: _____

Problem Solving

Solve each problem.

1. A rectangular sheet of paper measures $8\frac{1}{2}$ inches by 11 inches. What is the area of the sheet of paper?

The area is _____ square inches.

2. A square-shaped picture measures $5\frac{1}{2}$ inches along each edge. What is the area of the picture?

The area is _____ square inches.

3. Mr. Hudson's rectangular lot is 18 meters wide and 31 meters long. What is the area of his lot?

The area is _____ square meters.

4. Mr. Hatfield asked the pupils to draw a rectangle that has an area of 15 square inches. Mark drew a 3-inch by 5-inch rectangle. Millie drew a $2\frac{1}{2}$-inch by 6-inch rectangle. Were both pupils correct?

Were both pupils correct? _____

5. Which has the greater area, a rectangle 4 meters by 6 meters or a square that is 5 meters long on each side? How much greater?

The _____ has the greater area.

It is _____ square meter greater.

6. Trina had a rectangular-shaped board that measured 6.5 feet by 4.2 feet. A 2-foot piece was cut off the length. What is the area of the piece cut off? What is the area of the other part?

The area of the cut-off piece is _____ square feet.

The area of the other part is _____ square feet.

1.	2.
3.	4.
5.	
6.	

Perfect score: 8   My score: _____

144

Lesson 5   Area of Triangles

To find the *area measure* (A) of a triangle, take $\frac{1}{2}$ the product of the measures of the **base** (b) and the *height* (h).

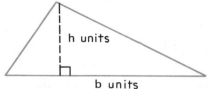

Find A if b=16 and h=5.

$$A = \tfrac{1}{2} \times b \times h$$

$$= \tfrac{1}{2} \times 16 \times 5$$

$$= \tfrac{1}{2} \times \underline{\ 80\ }$$

$$= \underline{\ 40\ }$$

The area is __40__ square units.

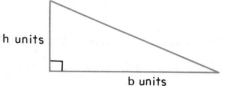

Find A if b=12 and h=7.

$$A = \tfrac{1}{2} \times b \times h$$

$$= \tfrac{1}{2} \times \underline{\qquad} \times \underline{\qquad}$$

$$= \underline{\qquad} \times \underline{\qquad}$$

$$= \underline{\qquad}$$

The area is _____ square units.

Find the area of each triangle below.

a	b	c

1.

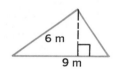 6 m / 9 m

3 ft / 4 ft

5 cm / 12 cm

_____ square meters   _____ square feet   _____ square centimeters

Complete the table for each triangle described below.

	base	height	area
2.	12 m	6 m	_____ square meters
3.	15 km	4 km	_____ square kilometers
4.	$3\frac{1}{2}$ in.	4 in.	_____ square inches
5.	7.5 cm	8.2 cm	_____ square centimeters
6.	$4\frac{1}{4}$ in.	$3\frac{1}{2}$ in.	_____ square inches
7.	12 yd	$14\frac{1}{3}$ yd	_____ square yards

Perfect score: 9   My score: _____

Problem Solving

Solve each problem.

1. The base of a triangle is 6 meters long. The height is 9 meters. What is the area of the triangle?

The area is _____ square meters.

2. Suppose both the base and height of a triangle are half the length of those in problem 1. What is the area of the triangle?

The area is _____ square feet.

3. A triangular playing field has a base of 42 feet and a height of 54 feet. What is the area of the field?

The area is _____ square feet.

4. A square photograph was cut into two parts as shown. What is the area of each part?

7 cm

The area is _____ square centimeters.

5. Suppose the colored region is removed from the drawing at the right. What is the area of the part removed? What is the area of the remaining part?

4 in.

4 in.   2. 4 in.

6 in.

The area of the part that is removed

is _____ square inches.

The area of the part that remains is

_____ square inches.

6. In a triangle the sides that form a right angle are 3.2 yards long and 1.6 yards long. What is the area of the triangle?

The area is _____ square yards.

1.	2.
3.	4.
5.	
6.	

Perfect score: 7   My score: _____

146

Lesson 6   Area of Circles

$A = \pi \times r \times r$ is used to find the area measure of a circle.

Find A if $r = 4$.
$$A = \pi \times r \times r$$

$$= 3.14 \times 4 \times \rule{1.5cm}{0.4pt}$$

$$= 3.14 \times \rule{1.5cm}{0.4pt}$$

$$= \rule{2cm}{0.4pt}$$

The area is _____ square units.

Find A if $d = 14$.
(Since $d = 14$, $r = d \div 2$ or 7.)
$$A = \pi \times r \times r$$

$$= \tfrac{22}{7} \times \rule{1.2cm}{0.4pt} \times \rule{1.2cm}{0.4pt}$$

$$= \tfrac{22}{7} \times \rule{1.5cm}{0.4pt}$$

$$= \rule{2cm}{0.4pt}$$

The area is _____ square units.

Find the area of each circle below. Use 3.14 for π.

| a | b | c |

1.

_____ square yards

_____ square centimeters

_____ square feet

2.

_____ square feet

_____ square meters

_____ square kilometers

Complete the table. Use $3\tfrac{1}{7}$ for π.

	diameter	radius	area
3.	_____ ft	21 ft	_____ square feet
4.	14 m	_____ m	_____ square meters
5.	_____ in.	16 in.	_____ square inches
6.	70 ft	_____ ft	_____ square feet
7.	_____ mm	28 mm	_____ square millimeters

Perfect score: 16   My score: _____

Problem Solving

Solve each problem. Use $3\frac{1}{7}$ for π in problems 1-4. Use 3.14 for π in problems 5-7.

1. What is the area of a circle if the diameter of the circle is 4 inches?

The area is _____ square inches.

2. What is the area of a circular flower bed if the radius is 42 inches?

The area is _____ square inches.

3. What would be the area of the flower bed in problem 2 if the radius were half as long?

The area would be _____ square inches.

4. What would be the area of the flower bed in problem 2 if the radius were twice as long?

The area would be _____ square inches.

5. A quarter has a radius of 12 millimeters. Find the area of a quarter.

The area is _____ square millimeters.

6. What is the area of the circle enclosed in the square? What is the area of the square? 6 ft

The area of the circle is _____ square feet.

The area of the square is _____ square feet.

7. Which has the greater area, circle **A** with a diameter of 4 meters or circle **B** with a radius of 3 meters? How much greater?

Circle _____ has the greater area.

It is _____ square meters greater.

1.	
2.	**3.**
4.	**5.**
6.	
7.	

Perfect score: 9   My score: _____

148

Lesson 7   Area

Study how the area of this figure is found.

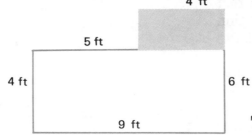

area of colored part
$A = l \times w$
$= 4 \times 2$

$= \underline{\hspace{1cm}}$

area of white part
$A = l \times w$
$= 9 \times 4$

$= \underline{\hspace{1cm}}$

The total is _____ + _____ or _____ square feet.

Find the area of each figure.

a

1.

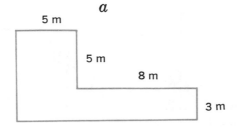

_____ square meters

b

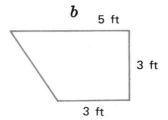

_____ square feet

2.

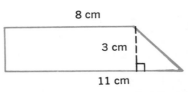

_____ square centimeters

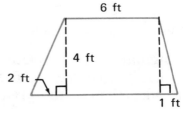

_____ square feet

3.

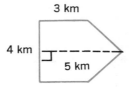

_____ square kilometers

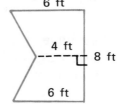

_____ square feet

Find the area of the colored part of each figure.

4.

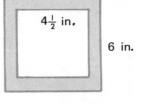

_____ square inches

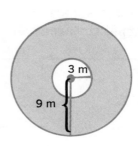

_____ square meters

Perfect score: 8   My score: _____

Problem Solving

Solve each problem. Use 3.14 for π.

1. Mrs. Roberts has a one-story house that is 32 feet wide and 48 feet long. She plans to build an addition 12 feet by 16 feet. What will the total area of the house be then?

The total area will be _____ square feet.

2. Mary has a garden shaped as shown below. What is the area of the garden?

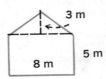

The area is _____ square meters.

3. A rectangular piece of cloth is 8 meters by 20 meters. A piece with an area of 144 square meters is removed. What is the area of the remaining piece?

The area is _____ square meters.

4. A rectangular piece of cloth is 7 meters by 8 meters. A square piece that is 5 meters along each side is removed. What is the area of the remaining piece?

The area is _____ square meters.

5. The Treadles have a lot that is 55 feet by 120 feet. Their house is 32 feet by 38 feet. Their garage is 20 feet by 24 feet. How much of their lot is vacant?

_____ square feet of their lot is vacant.

6. How many square feet of carpeting would be needed to carpet the area shown below?

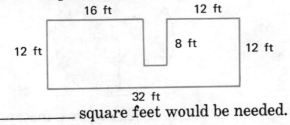

_____ square feet would be needed.

1.
2.
3.
4.
5.
6.

Perfect score: 6   My score: _____

CHAPTER 10 TEST

Complete the table for each rectangle described below.

	length	width	perimeter	area
1.	7 m	5 m	_____ m	_____ square meters
2.	3 in.	1.5 in.	_____ in.	_____ square inches
3.	$3\frac{1}{2}$ in.	$1\frac{1}{2}$ in.	_____ in.	_____ square inches
4.	5 cm	5 cm	_____ cm	_____ square centimeters

Complete the table for each triangle described below.

	base	height	area
5.	7 cm	8 cm	_____ square centimeters
6.	$9\frac{1}{2}$ ft	4 ft	_____ square feet
7.	7.5 yd	3.4 yd	_____ square yards

Complete the table for each circle described below. Use 3.14 for π.

	diameter	radius	circumference	area
8.	8 m	_____ m	_____ m	_____ square meters
9.	_____ ft	100 ft	_____ ft	_____ square feet
10.	_____ cm	_____ cm	18.84 cm	_____ square centimeters

Perfect score: 20   My score: _____

151

PRE-TEST—Volume

Find the volume of each rectangular solid.

| a | b | c |

1.

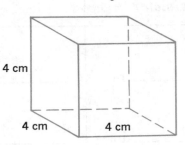

2 m

3 m

6 m

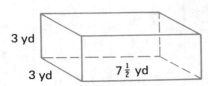

4 cm

4 cm   4 cm

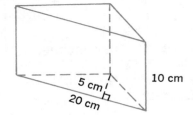

3 yd

3 yd $7\frac{1}{2}$ yd

_____ cubic meters   _____ cubic centimeters   _____ cubic yards

Find the volume of each triangular prism.

2.

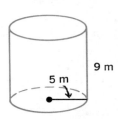

5 cm

20 cm   10 cm

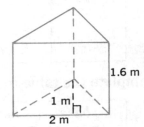

3 in.   4 in.   6 in.

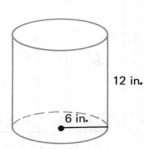

1 m   2 m   1.6 m

_____ cubic centimeters   _____ cubic inches   _____ cubic meters

Find the volume of each cylinder. Use 3.14 for π.

3.

5 m   9 m

8 cm   10 cm

6 in.   12 in.

_____ cubic meters   _____ cubic centimeters   _____ cubic inches

Find the volume of each.

4.

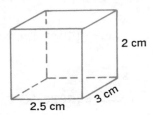

2 cm

2.5 cm   3 cm

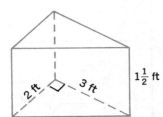

2 ft   3 ft $1\frac{1}{2}$ ft

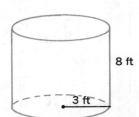

3 ft   8 ft

_____ cubic centimeters   _____ cubic feet   _____ cubic feet

Perfect score: 12   My score: _____

Lesson 1 Volume of Rectangular Solids

The *volume measure* (V) *of a rectangular solid* is the product of the *area measure of the base* (B) and the measure of the *height* (h). $V = B \times h$

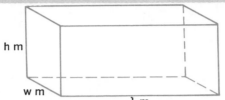

Find V if $l = 6\frac{1}{2}$, $w = 4$, and $h = 3$.

$$V = \overbrace{B}^{} \times h$$
$$= l \times w \times h$$
$$= 6\frac{1}{2} \times 4 \times 3$$
$$= \underline{\hspace{1cm}} \times 3$$
$$= \underline{\hspace{1cm}}$$

The volume is _____ cubic meters.

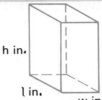

Find V if $l = 4$, $w = 3$, and $h = 6$.

$$V = B \times h$$
$$= l \times w \times h$$
$$= \underline{\hspace{0.5cm}} \times \underline{\hspace{0.5cm}} \times \underline{\hspace{0.5cm}}$$
$$= \underline{\hspace{1cm}} \times \underline{\hspace{0.5cm}}$$
$$= \underline{\hspace{1cm}}$$

The volume is _____ cubic inches.

Find the volume of each rectangular solid.

a   b   c

1.

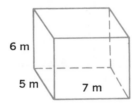

6 m   5 m   7 m

3 in.   3 in.   3 in.

5.5 cm   3 cm   3 cm

_____ cubic meters   _____ cubic inches   _____ cubic centimeters

Complete the table for each rectangular solid described below.

	length	*width*	*height*	*volume*
2.	8 m	6 m	3 m	_____ cubic meters
3.	$7\frac{1}{2}$ ft	4 ft	6 ft	_____ cubic feet
4.	3.2 cm	3.2 cm	4 cm	_____ cubic centimeters
5.	$5\frac{1}{4}$ yd	$2\frac{1}{4}$ yd	3 yd	_____ cubic yards
6.	$3\frac{1}{2}$ ft	$2\frac{1}{2}$ ft	1 ft	_____ cubic feet

Perfect score: 8   My score: _____

Problem Solving

Solve each problem.

1. The bottom of a box is $10\frac{1}{2}$ inches long and 5 inches wide. The box is **3** inches high. What is the volume of the box?

The volume is _____ cubic inches.

2. A box is 12 centimeters wide, 18 centimeters long, and 6 centimeters deep. What is its volume?

The volume is _____ cubic centimeters.

3. Assume each dimension in problem **2** is doubled. What would the volume of the drawer be?

It would be _____ cubic centimeters.

4. A cube with each dimension 12 inches has a volume of 1 cubic foot. How many cubic inches are in 1 cubic foot?

_____ cubic inches are in 1 cubic foot.

5. A cube with each dimension 3 feet has a volume of 1 cubic yard. How many cubic feet are in 1 cubic yard?

_____ cubic feet are in 1 cubic yard.

6. Mark has a box that is 3 feet high, 4 feet wide, and 5 feet long. Jeff has a box that measures 4 feet along each edge. Whose box has the greater volume? How much greater is it?

_____ box has the greater volume.

It is _____ cubic feet greater.

7. Anne has 500 cubes, each with edges 1 centimeter long. How many more cubes does she need to fill the box?

She needs _____ more cubes.

7 cm

8 cm

10 cm

1.	
2.	**3.**
4.	**5.**
6.	
7.	

Perfect score: 8   My score: _____

154

Lesson 2   Volume of Triangular Prisms

The *volume measure* (*V*) *of a triangular prism* is the product of the *area measure of the base* (*B*) and the measure of the *height* (*h*). $V = B \times h$

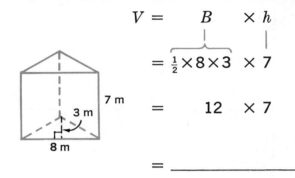

$$V = \underbrace{B}_{= \frac{1}{2} \times 8 \times 3} \times \overset{\mid}{\underset{7}{h}}$$

$= \frac{1}{2} \times 8 \times 3 \times 7$

$= \qquad 12 \quad \times 7$

$= \underline{\hspace{2cm}}$

The volume is _____ cubic meters.

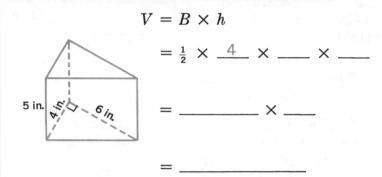

$V = B \times h$

$= \frac{1}{2} \times \underline{4} \times \underline{\hspace{1cm}} \times \underline{\hspace{1cm}}$

$= \underline{\hspace{2cm}} \times \underline{\hspace{1cm}}$

$= \underline{\hspace{2cm}}$

The volume is _____ cubic inches.

Find the volume of each triangular prism.

| *a* | *b* | *c* |

1.

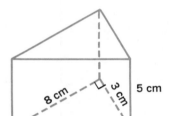

_____ cubic centimeters

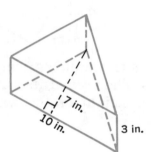

_____ cubic inches

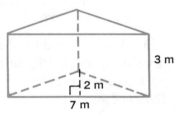

_____ cubic meters

2.

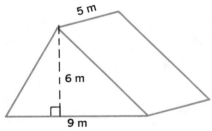

_____ cubic meters

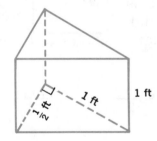

_____ cubic foot

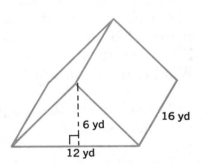

_____ cubic yards

Perfect score: 6   My score: _____

Problem Solving

Solve each problem.

1. Find the volume of the rectangular solid shown at the right.

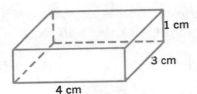

The volume is _____ cubic meters.

2. The rectangular solid in problem **1** was cut to form these two triangular prisms. What is the volume of each triangular prism?

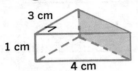 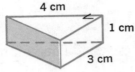

The volume is _____ cubic meters.

3. The tent is shaped like a triangular prism. Find its volume.

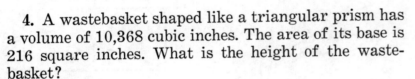

The volume is _____ cubic meters.

4. A wastebasket shaped like a triangular prism has a volume of 10,368 cubic inches. The area of its base is 216 square inches. What is the height of the wastebasket?

The height is _____ inches.

5. The top part of the tent is shaped like a triangular prism. Find the volume of the top part. The bottom part of the tent is shaped like a rectangular solid. Find the volume of the bottom part. What is the volume of the tent?

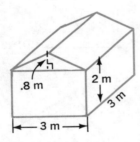

The volume of the top part is _____ cubic meters.

The volume of the bottom part is _____ cubic meters.

The volume of the tent is _____ cubic meters.

1.

2.

3.   **4.**

5.

Perfect score: 7   My score: _____

156

Lesson 3   Volume of Cylinders

The *volume measure (V) of a cylinder* is the product of the *area measure of the base* (B) and the measure of the *height (h)*. $V = B \times h$

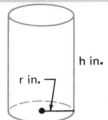

Find V if $r=2$ and $h=6$.

$$V = \underbrace{B}_{} \times h$$
$$= \pi \times r \times r \times h$$
$$= \tfrac{22}{7} \times 2 \times 2 \times 6$$
$$= \tfrac{528}{7}$$
$$= 75\tfrac{3}{7}$$

The volume is ___$75\frac{3}{7}$___ cubic inches.

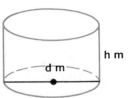

Find V if $d=8$ and $h=5$.
(Since $d=8$, $r=4$.)

$$V = B \times h$$
$$= \pi \times r \times r \times h$$
$$= 3.14 \times \underline{\quad} \times \underline{\quad} \times \underline{\quad}$$
$$= \underline{\qquad\qquad}$$

The volume is _____ cubic meters.

Find the volume of each cylinder. Use $3\frac{1}{7}$ for π.

a	b	c

1.

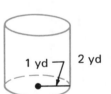

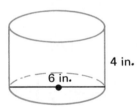

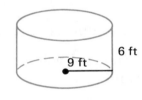

_____ cubic yards   _____ cubic inches   _____ cubic feet

Complete the table for each cylinder described below. Use 3.14 for π.

	diameter	radius	height	volume
2.	___ m	3 m	6 m	_____ cubic meters
3.	___ cm	7 cm	4 cm	_____ cubic centimeters
4.	10 yd	___ yd	3 yd	_____ cubic yards
5.	___ cm	5 cm	2 cm	_____ cubic centimeters
6.	18 m	___ m	3 m	_____ cubic meters

Perfect score: 13   My score: _____

Problem Solving

Solve each problem. Use 3.14 for π.

1. A coffee can has a 3-inch radius and is 8 inches high. What is the volume of the coffee can?

The volume is _____ cubic inches.

2. Suppose the can in problem **1** is $\frac{1}{2}$ filled with coffee. How many cubic inches of coffee are in the can?

There are _____ cubic inches of coffee.

3. A cylindrical container has a diameter of 4 inches and is 12 inches high. What is its volume?

Its volume is _____ cubic inches.

4. Suppose the container in problem **3** had a diameter of 12 inches and was 4 inches high. What would its volume be?

Its volume would be _____ cubic inches.

5. A cylindrical tank has a radius of 8 meters and is 4 meters high. What is the volume of the tank?

The volume is _____ cubic meters.

6. Find the volume of the wading pool.

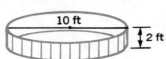

The volume is _____ cubic feet.

7. Which can has the greater volume? How much greater?

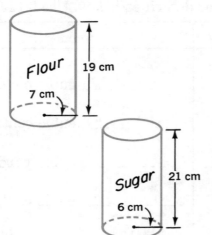

The _____ can has the greater volume.

It is _____ cubic centimeters greater.

1.	2.
3.	4.
5.	6.
7.	

Perfect score: 8   My score: _____

CHAPTER 11 TEST

Complete the table for each rectangular solid described below.

	length	width	height	volume
1.	6 m	4 m	8 m	_____ cubic meters
2.	7 cm	7 cm	7 cm	_____ cubic centimeters
3.	9 ft	$7\frac{1}{2}$ ft	6 ft	_____ cubic feet
4.	$5\frac{1}{2}$ in.	$3\frac{1}{2}$ in.	2 in.	_____ cubic inches

Find the volume of each triangular prism.

a	b	c

5.

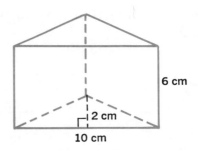   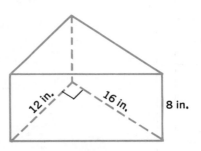   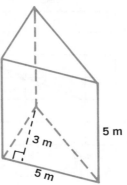

_____ cubic centimeters   _____ cubic inches   _____ cubic meters

Complete the table for each cylinder described below. Use 3.14 for π.

	diameter	radius	height	volume
6.	_____ ft	12 ft	8 ft	_____ cubic feet
7.	16 cm	_____ cm	19 cm	_____ cubic centimeters
8.	_____ yd	10 yd	12 yd	_____ cubic yards
9.	21 m	_____ m	14 m	_____ cubic meters

Perfect score: 15   My score: _____

PRE-TEST   Graphs and Probability

Use the graph at the right to answer each question.

1. Which salesperson had the greatest amount of sales? _____

2. Which salesperson had the least amount of sales? _____

3. Who had sales of about $50,000? _____

4. About what is the total amount of sales for Ed and Meg? _____

5. About what is the difference between Al's sales and Jan's sales? _____

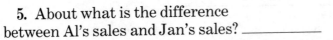

Sales Report

Use the graph at the left to answer the following question.

6. List at least one thing that is wrong with the graph at the left.

_____

_____

_____

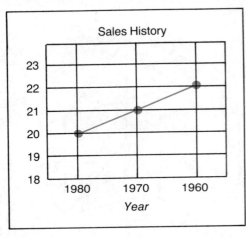

To choose a partner, you draw one of the cards at the right. Write the probability in simplest form that you will pick:

7. Mr. Alvers _____

8. A man _____

9. Mr. or Mrs. Von _____

10. a person who has

a first name of Pat _____

11. a person who does **not**

have a first name of Ken _____

Mrs. Pat Von	Mr. Ken Alvers
Mr. Ken Bruns	Ms. Kari Yoshita
Mr. Pat Wittier	Mr. Ken Von

Perfect score: 11   My score: _____

Lesson 1 Bar Graphs

Bar graphs are used to compare amounts.

Horizontal Bar Graph

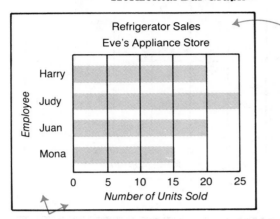

The **title** tells what the graph is about.

The **number scale** should start at 0.

Vertical Bar Graph

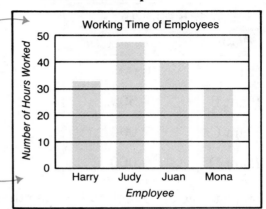

The **labels** identify parts of the graph. There is a bar for each employee.

Use the bar graphs above to answer each question.

1. What is the difference between a horizontal and a vertical bar graph?

The difference is _____.

2. How many units did Harry sell?

Harry sold _____ units.

3. Who sold the most units?

_____ sold the most units.

4. Who sold the fewest units?

_____ sold the fewest units.

5. Who worked the most hours?

_____ worked the most hours.

6. How many more hours did Juan work than Mona?

Juan worked _____ hours more than Mona.

Perfect score: 6   My score: _____

161

Lesson 2   Bar Graphs

You can make a bar graph from data.

Units Manufactured	
Plant	*Number of Units*
Plant A	6,000
Plant B	4,500
Plant C	2,000
Plant D	5,000

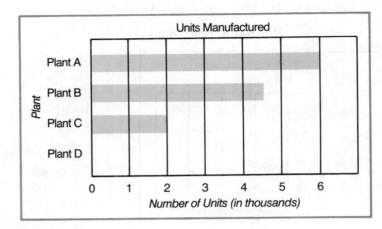

Complete the bar graph by drawing the bar for Plant D.

Use the information in each table to help you complete each bar graph.

1.

Committee Election Results	
Candidate	*Number of Votes*
Pat	8
Jerri	6
Kami	12
Lina	9
David	10
Maria	12

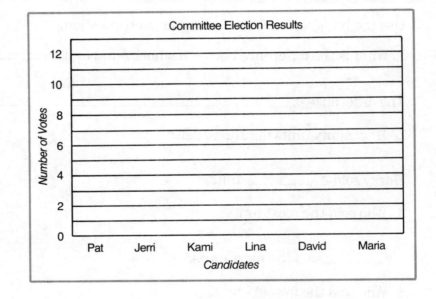

2.

Cost of Parts	
Part	*Cost*
Part A	$1.50
Part B	$3.00
Part C	$1.75
Part D	$2.25

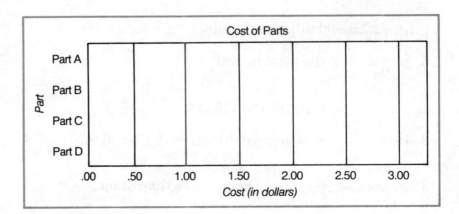

Perfect score: 10   My score: _____

Lesson 3 Line Graphs

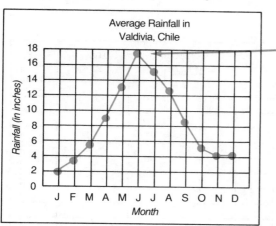

Line graphs make it easy to see change.

June has the most rainfall, about 17.5 inches.

The scale cannot be read exactly, so you must **estimate** each amount.

In what month does Valdivia get the least

rainfall? _____

What is the average rainfall for August in

Valdivia? _____

Use the line graphs to help you answer each question.

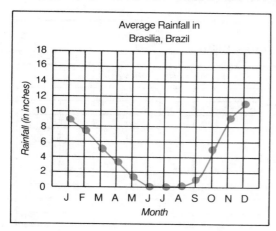

1. During which month does Brasilia have the

greatest average rainfall? _____

2. During which months does Brasilia get no

rainfall? _____

3. During which month does Brasilia get about

7.5 inches of rainfall? _____

4. What is the average rainfall in Brasilia in

October? _____

5. In which month was the greatest number of

compact disks sold? _____

The least number? _____

6. In which months were the sales about the

same? _____

7. Did the number of sales increase or decrease
during the first three months of the year?

_____

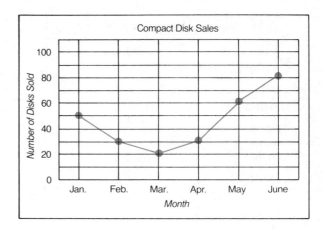

Perfect score: 8   My score: _____

Lesson 4 Making Line Graphs

You can make a line graph from data.

Computer System Sales	
Month	*Systems Sold*
Jan.	38
Feb.	30
Mar.	22
Apr.	16
May	26
June	30

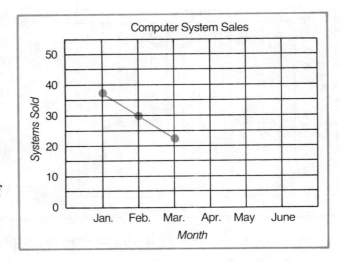

The dot ● for Jan. is drawn to show the number of systems sold was _____.

Complete the line graph for April, May, and June.

Use the information in each table to help you complete each line graph.

1.

Weight Record	
Month	*Weight (in pounds)*
1	150
2	140
3	135
4	125
5	120
6	125

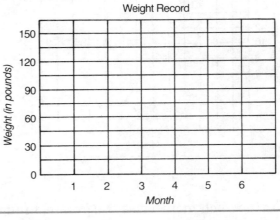

2.

Average Rainfall in Denton, Texas	
Month	*Rainfall (in inches)*
January	1.6
February	2.4
March	2.6
April	4.2
May	4.9
June	3.3
July	1.9
August	2.0
September	3.8
October	2.5
November	2.2
December	2.0

Perfect score: 18   My score: _____

Lesson 5   Misleading Graphs

Graphs make information easy to read. But graphs can also be drawn so they create an incorrect impression.

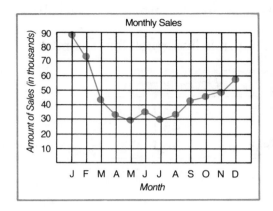

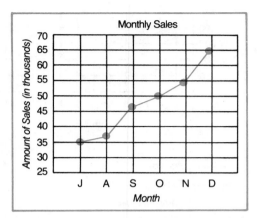

By not starting the scale at 0, it appears that sales are growing very fast.

By only showing part of the year's sales, the graph is very misleading!

Use the graphs to help you answer each question.

1.

←**Graph A**

Graph B→

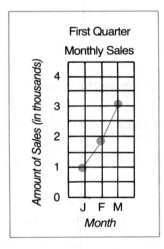

What has been done to make sales look better in Graph B than in Graph A?

2. How many new cars were sold in March? _____

3. How many new cars were sold in April? _____

4. How many new cars were sold in January? _____

5. Are sales increasing or decreasing? _____

6. How is this graph misleading?

Graph C

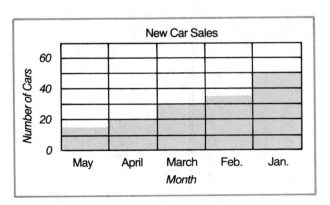

Perfect score: 6   My score: _____

165

Lesson 6   Multiple Line Graphs

Multiple line graphs make it easy to see change and to compare numbers.

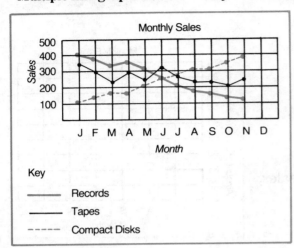

Notice a different kind of line is used for each type of item sold. The **key** shows what each line represents.

Complete the multiple line graph for December.

Item	December Sales
records	180
tapes	450
compact disks	360

Use the line graph above to help you answer each question.

1. How many records were sold in January? _____

2. How many tapes were sold in February? _____

3. How many compact disks were sold in May? _____

4. Are record sales increasing or decreasing? _____

5. Are compact disk sales increasing or decreasing? _____

Use the information in the chart below to complete a double line graph.

6. A company kept track of the following information for 10 days.

Day	Number of Workers Absent	Number of Units Produced
1	8	18
2	15	14
3	5	20
4	10	15
5	7	17
6	0	25
7	2	22
8	8	17
9	15	15
10	20	12

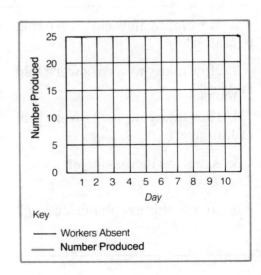

Perfect score: 16   My score: _____

166

Lesson 7   Probability

You draw one of the cards shown at the right without looking. You would like to know your *chance* or **probability** of getting a card that says *win*.

Each card (possible result) is called an **outcome**. There are 10 cards. There are 10 possible outcomes. Since you have the same chance of drawing any of the cards, the outcomes are **equally likely**.

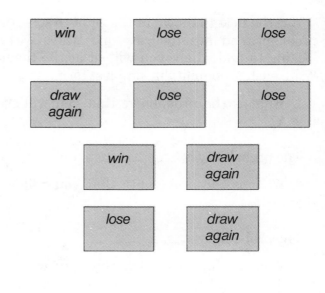

number of outcomes
that say win

$\frac{2}{10}$ or $\frac{1}{5}$ Write the probability in simplest form.

number of
possible outcomes

The probability of drawing a card that says *win* is $\frac{1}{5}$.

You spin the wheel shown at the right. Find the probability of the spinner stopping on:

1. a blue letter _____

2. the letter A _____

3. a black letter _____

4. the letter C _____

5. a black letter B _____

6. the letter D or B _____

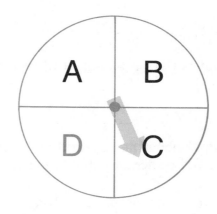

You pick a marble without looking. In simplest form, what is the probability of picking:

7. white _____

8. black _____

9. blue _____

10. a marble that is **not** white _____

Perfect score: 10   My score: _____

167

Problem Solving

Use the tickets shown at the right to solve each problem. You are to draw one ticket without looking to find where you will sit at the theater. Write each probability in simplest form.

1. What is the probability that you will sit in row A?

The probability is _____.

2. What is the probability that you will sit in seat 3?

The probability is _____.

3. What is the probability that you will sit in row B?

The probability is _____.

4. What is the probability that you will sit in seat 1?

The probability is _____.

5. What is the probability that you will **not** sit in seat 1?

The probability is _____.

Row A Seat 1	Row A Seat 2
Row A Seat 3	Row A Seat 4
Row A Seat 5	Row A Seat 6
Row B Seat 1	Row B Seat 2
Row B Seat 3	Row C Seat 1
Row C Seat 2	Row D Seat 1

The sandwiches at the right are at a picnic. You pick one sandwich without looking. Solve each problem. Write each probability in simplest form.

6. What is the probability that you will pick roast beef?

The probability is _____.

7. What is the probability that you will pick cheese?

The probability is _____.

8. What is the probability that you will **not** pick cheese?

The probability is _____.

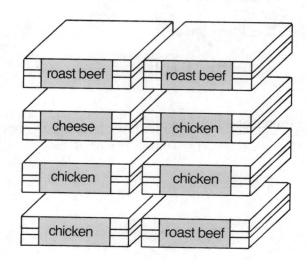

Perfect score: 8   My score: _____

168

Lesson 8   0 and 1 Probabilities

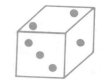

The faces of a die have
1, 2, 3, 4, 5, and 6 dots.
You roll one die one time.

probability of rolling a 2	probability of rolling a number less than 7	probability of rolling a 7
$\frac{1}{6}$	$\frac{6}{6}$ or 1	$\frac{0}{6}$ or 0

A probability of 1 means the outcome is **certain** to happen.

A probability of 0 means the outcome will **never** happen.

To decide which committee you will be on, you are to draw one of the slips shown at the right without looking. These are the only committees you can be on. Write each probability in simplest terms.

refreshments	cleanup
refreshments	cleanup
refreshments	cleanup
decorations	cleanup

1. How many slips are there? _____

2. What is the probability of being on the cleanup committee? _____

3. What is the probability of being on the invitation committee? _____

4. What is the probability of being on the decoration committee? _____

5. What is the probability of **not** being on the decoration committee? _____

6. What is the sum (total) of the probabilities in exercises 4 and 5? _____

7. What is the probability of being on the refreshments committee? _____

8. When you add the probability that an outcome **will** happen and the probability that it **will not** happen, the answer is _____ .

Perfect score: 8   My score: _____

Problem Solving

Solve each problem. Write each probability in simplest form.

1. You are taking a multiple-choice test. Each item has 4 choices. You have no idea which is the correct choice. What is the probability that you will guess the correct choice?

The probability is _____.

2. Suppose that each item on the test in 1 had 5 choices. You still have no idea which is the correct choice. What is the probability that you will guess the correct choice?

The probability is _____.

3. You draw 1 marble from a bag containing 6 marbles. There are 4 white marbles and 2 black marbles. What is the probability that you will draw a red marble?

The probability is _____.

4. You pick one of the letter cards shown at the right without looking. What is the probability that you will pick a vowel (a, e, i, o, u)?

The probability is _____.

5. You pick one of the letter cards shown at the right without looking. What is the probability that the letter on the card is in the word *CALIFORNIA*?

The probability is _____.

6. You pick one of the letter cards shown at the right without looking. What is the probability that the letter on the card is **not** in the word *CALIFORNIA*?

The probability is _____.

7. You pick one of the number cards shown at the right without looking. What is the probability that you will pick a number greater than 4?

The probability is _____.

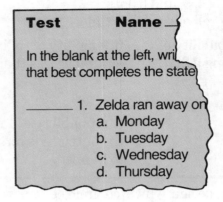

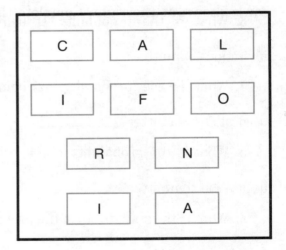

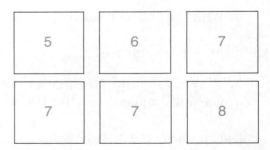

Perfect score: 7   My score: _____

170

Lesson 9 Sample Spaces

Suppose you have a choice of jewelry. You can have a bracelet, a necklace, or a pin. The jewelry can be gold, silver, or platinum.

You can show all the different outcomes in a table like the one shown at the right.

A list or a table of all the possible outcomes is called a **sample space**.

Jewelry Piece

Color		bracelet (B)	necklace (N)	pin (P)
	gold (g)	Bg	Ng	Pg
	silver (s)	Bs	Ns	Ps
	platinum (p)	Bp	Np	Pp

Use the sample space above to answer each question.

1. How many possible outcomes are there?

_____

2. Suppose your jeweler chooses a combination for you at random.

What is the probability of your getting a gold

necklace? _____

Without looking, you pick marbles from two bowls shown at the right. Complete the sample space to show the outcomes. Then use the table to answer each question. Write each probability in simplest form.

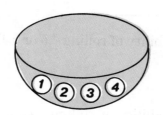 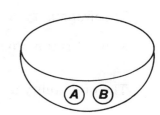

Blue Bowl

White Bowl	1	2	3	4
A	1A	2A		
B	1B			

3. How many outcomes are there?   _____

4. What is the probability of drawing a 2 and an A?   _____

5. What is the probability of drawing a B and an even number? _____

6. What is the probability of drawing any number and a B?   _____

Perfect score: 11   My score: _____

171

Problem Solving

Complete the sample space below to show all the possible outcomes of rolling a blue die and a black die. Then use the sample space to solve each problem. Write each probability in simplest form.

Blue Die

	1	2	3	4	5	6
1	1,1	2,1	3,1	4,1		
2	1,2	2,2	3,2			
3	1,3					
4	1,4					
5						
6						

Black Die

1. What is the probability of rolling a 1 and a 3?

The probability is _____.

2. What is the probability of rolling 4's on both dice?

The probability is _____.

3. What is the probability of rolling 4,5 or 5,4?

The probability is _____.

4. What is the probability of rolling two dice that total 10?

The probability is _____.

5. What is the probability of rolling two dice that total 20?

The probability is _____.

6. What is the probability of rolling two dice that total less than 13?

The probability is _____.

Perfect score: 33   My score: _____

172

Lesson 10   Problem Solving

A local store is running a contest. Each time you enter the store you get a ticket. The sign below shows the probability of winning a prize. Use the sign below to solve each problem. Write each probability in simplest form.

1. What is the probability of winning the grand prize?

The probability is _____.

2. What is the probability of winning the second prize?

The probability is _____.

3. What is the probability of winning the fifth prize?

The probability is _____.

4. What is the probability of not winning any prize?

The probability is _____.

BRUNS' SHOE STORE	
Chances of winning:	
Prize	*Number of Tickets*
Grand prize	1
Second prize	5
Third prize	10
Fourth prize	20
Fifth prize	25
Total number of tickets to be given out: 1,000	

One person will be selected at random to represent the company at a special event. The table below shows how many people volunteered to be selected. Each name is written on a slip of paper, and one slip is to be drawn. Use the table to solve each problem. Write each probability in simplest form.

5. What is the probability that a woman will be selected?

The probability is _____.

6. What is the probability that a person from Department C will be selected?

The probability is _____.

7. What is the probability that a woman from Department A will be selected?

The probability is _____.

Department	*Women*	*Men*
A	0	5
B	6	1
C	2	10
D	12	4

Perfect score: 7   My score: _____

173

CHAPTER 12 TEST

Use the graph *at the right* to answer each question.

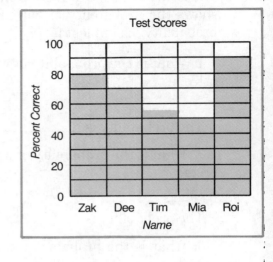

1. Which person had the highest test score?

   _____

2. Which person had the lowest test score?

   _____

3. What is the difference between Zak's and

   Dee's scores? _____

4. If a score of 60% is passing, how many people

   passed the test? _____

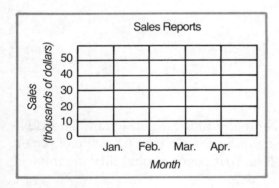

Complete the line graph at the left, using the following information.

5.

Sales Reports	
Month	*Sales* ($)
January	$30,000
February	50,000
March	25,000
April	43,000

You roll a die with faces marked as shown at the right. Write the probability in simplest form that you will roll:

6. ahead 4 _____

7. back 3 _____

8. ahead any number _____

9. a face with an odd number _____

10. a face that does not say back _____

ahead 4	back 2	ahead 3

lose a turn	back 1	ahead 2

Perfect score: 10   My score: _____

NAME _____

Complete.

	a	b	c	d	e

1.
$$\begin{array}{r} 1\ 4\ 7\ 5\ 6 \\ 6\ 4\ 5\ 3 \\ +7\ 2\ 1\ 4\ 1 \\ \hline \end{array}$$
$$\begin{array}{r} 8\ 3\ 4 \\ 1\ 7\ 8 \\ 2\ 4\ 8 \\ +4\ 3 \\ \hline \end{array}$$
$$\begin{array}{r} 7\ 8\ 6 \\ -6\ 9\ 7 \\ \hline \end{array}$$
$$\begin{array}{r} 3\ 5\ 0\ 0\ 6 \\ -1\ 3\ 5\ 1\ 7 \\ \hline \end{array}$$
$$\begin{array}{r} 4\ 5\ 6\ 8 \\ -2\ 5\ 7\ 9 \\ \hline \end{array}$$

2.
$$\begin{array}{r} 7\ 5\ 6 \\ \times 4\ 0 \\ \hline \end{array}$$
$$\begin{array}{r} 8\ 2\ 6\ 7 \\ \times 3\ 5\ 6 \\ \hline \end{array}$$
$7\overline{)9\ 5}$
$9\overline{)1\ 8\ 6\ 2}$
$56\overline{)8\ 7\ 8\ 2}$

3.
$$\begin{array}{r} \$4\ 2.8\ 3 \\ 1.9\ 5 \\ +2.7\ 8 \\ \hline \end{array}$$
$$\begin{array}{r} .1\ 7\ 8\ 3 \\ +1.6\ 2 \\ \hline \end{array}$$
$$\begin{array}{r} .8 \\ -.4\ 3\ 2 \\ \hline \end{array}$$
$$\begin{array}{r} \$1\ 8.2\ 4 \\ -6.8\ 9 \\ \hline \end{array}$$
$$\begin{array}{r} 2\ 7.0\ 8 \\ -1\ 6.9\ 0\ 7 \\ \hline \end{array}$$

4.
$$\begin{array}{r} .3\ 2\ 4 \\ \times 1.5 \\ \hline \end{array}$$
$$\begin{array}{r} 1\ 4.2\ 4 \\ \times 4.0\ 7 \\ \hline \end{array}$$
$8\overline{)1.6\ 0\ 8}$
$.06\overline{)1\ 9.2}$
$.45\overline{).6\ 8\ 4}$

5.
$\dfrac{2}{3}\times\dfrac{6}{7}$
$1\dfrac{1}{2}\times 2\dfrac{1}{3}$
$\dfrac{7}{8}\div 7$
$\dfrac{3}{10}\div\dfrac{2}{5}$
$8\dfrac{1}{8}\div 3\dfrac{3}{4}$

6.
$$\begin{array}{r} \dfrac{5}{6} \\ +\dfrac{7}{8} \\ \hline \end{array}$$
$$\begin{array}{r} 3\dfrac{9}{10} \\ +1\dfrac{2}{3} \\ \hline \end{array}$$
$$\begin{array}{r} \dfrac{8}{9} \\ -\dfrac{2}{3} \\ \hline \end{array}$$
$$\begin{array}{r} 1\dfrac{7}{8} \\ -\dfrac{3}{4} \\ \hline \end{array}$$
$$\begin{array}{r} 16\dfrac{3}{4} \\ -8\dfrac{9}{10} \\ \hline \end{array}$$

Continued on the next page.

Test—Chapters 1–5 (Continued)

Solve each of the following.

	a	b	c	d
7.	$\dfrac{6}{7}=\dfrac{n}{28}$	$\dfrac{3}{15}=\dfrac{8}{n}$	$\dfrac{n}{55}=\dfrac{4}{22}$	$\dfrac{1}{n}=\dfrac{80}{160}$
8.	$\dfrac{8}{n}=\dfrac{20}{15}$	$\dfrac{n}{25}=\dfrac{64}{100}$	$\dfrac{19}{20}=\dfrac{n}{100}$	$\dfrac{6}{100}=\dfrac{15}{n}$

Complete. Write each fraction in simplest terms.

a

	percent	fraction
9.	10%	
10.		$\frac{1}{2}$
11.	85%	
12.		$\frac{7}{8}$

b

	percent	decimal
	6%	
		.286
	$105\frac{1}{2}\%$	
		.085

Solve each problem.

13. A runner ran 3 miles in 15 minutes. At that rate, how long will it take to run 4 miles?

It will take _____ minutes.

14. Jack bought items for $2.39, $18.50, and $12.78. What is the total of those items?

The total is $_____.

15. A store is having a sale. All coats are marked $\frac{1}{4}$ off the original price. What is the percent of the discount on the coats?

The percent of discount is _____%.

16. Four plants cost $6. At that price, how much will 6 plants cost?

Six plants will cost $_____.

13.

14.

15.

16.

Perfect score: 50   My score: _____

FINAL TEST—Chapters 1–12

Complete.

	a	*b*	*c*	*d*	*e*
1.	2 5 4 2 6 5 6 4 5 3 + 2 2 4 4 1	6 2 0 2 7 6 + 2 5 7	9 8 0 − 5 9 7	9 4 0 2 0 − 2 4 8 7 3	4 6 5 × 4 9

2.　　 5 4 0 4 　　 8⟌9 3 5 2 　　 29⟌2 5 3 1 3 　　　 $1.6 9 　　 $1 3 7.8 9
　　　 × 9 3 6 　　　　　　　　　　　　　　　　　　　 + 9.3 6 　　 − 1 1 9.9 9

3.　　　 .2 1 8 　　 9 4.0 3 　　 8⟌1.6 0 8 　　 .06⟌1 9.2 　　 .45⟌.6 8 4
　　　　 × 1.8 　　　 × 7.0 4

	a	*b*	*c*	*d*
4.	$\frac{3}{5} \times \frac{7}{9}$	$4\frac{3}{8} \times 1\frac{3}{5}$	$\frac{6}{7} \div \frac{2}{3}$	$7\frac{1}{2} \div 3\frac{3}{4}$

5.　　 $\frac{2}{5}$ 　　　　　 $7\frac{1}{2}$ 　　　　　 $\frac{9}{10}$ 　　　　　 $6\frac{1}{4}$

　　 $+\frac{6}{7}$ 　　　 $+1\frac{2}{3}$ 　　　 $-\frac{2}{3}$ 　　　 $-3\frac{7}{8}$

Solve each of the following.

	a	*b*	*c*	*d*
6.	$\frac{n}{3} = \frac{4}{6}$	$\frac{9}{n} = \frac{27}{81}$	$\frac{7}{9} = \frac{70}{n}$	$\frac{8}{24} = \frac{n}{18}$

| 7. | $\frac{18}{25} = \frac{90}{n}$ | $\frac{8}{10} = \frac{n}{100}$ | $\frac{9}{n} = \frac{75}{100}$ | $\frac{n}{150} = \frac{6}{100}$ |

Continued on the next page.

Final Test (Continued)

Complete. Write each fraction in simplest terms.

	percent	fraction	decimal
8.	18%		
9.		$\frac{1}{4}$	
10.			.07

a

	percent	fraction	decimal
8.	6%		
9.		$\frac{1}{20}$	
10.			1.1

b

Complete the following.

	a	*b*	*c*
11.	_____ is 40% of 80.	25 is _____ % of 50.	_____ is 100% of 2.75.
12.	20 is 10% of _____.	18 is _____ % of 180.	300 is 5% of _____.
13.	2.2 is _____ % of 8.	_____ is 7.3% of 50.	63 is 90% of _____.
14.	72 is _____ % of 80.	32 is 4% of _____.	_____ is $1\frac{1}{2}$% of 30.

	principal	interest rate	time	interest	total amount
				a	*b*
15.	$800	8%	1 year		
16.	$1000	$10\frac{1}{2}$%	2 years		
17.	$780	12%	$3\frac{1}{4}$ years		
18.	$900	$7\frac{1}{2}$%	180 days		
19.	$2000	15%	90 days		

	a	*b*	*c*
20.	190 km = _____ m	80 mm = _____ cm	6 m = _____ cm
21.	8 liters = _____ ml	.9 kl = _____ liters	75 liters = _____ kl
22.	28 kg = _____ g	750 mg = _____ g	2 t = _____ kg

Continued on the next page.

Final Test (Continued)

Match each figure with its name. You will not use all the letters.

23. _____

24. _____

25. _____

26. _____

27. _____

28. _____

29. _____

30. _____

31. _____

32. _____

a. acute angle

b. acute triangle

c. isosceles triangle

d. line AB

e. line segment AB

f. obtuse angle

g. parallel lines

h. perpendicular lines

i. ray AB

j. ray BA

k. rectangle

l. rhombus

m. right triangle

n. square

Complete each of the following.

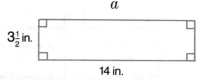

	square inches	square meters	square feet
33. **Area**	square inches	square meters	square feet
34. *Perimeter/ Circumference*	inches	meters	feet

Continued on the next page.

Final Test (Continued)

Find the volume of each. Use 3.14 for π.

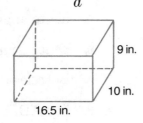 a

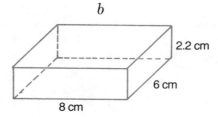 b

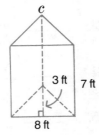 c

35.

_____ cubic inches

_____ cubic centimeters

_____ cubic feet

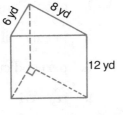

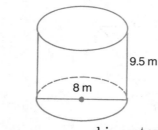

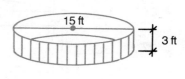

36.

_____ cubic yards

_____ cubic meters

_____ cubic feet

Solve each problem.

37. According to the graph at the right, how many parts did Department A make?

Department A made _____ parts.

38. According to the graph at the right, which Department made the most parts?

Department _____ made the most parts.

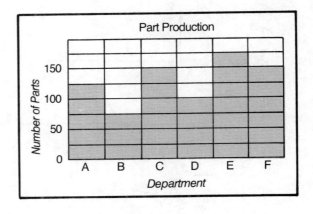

39. You are to choose one of the cards at the right without looking. What is the probability that you will draw a number less than 20?

The probability is _____.

40. You are to choose one of the cards at the right without looking. What is the probability that you will draw an even number?

The probability is _____.

10	50
12	40
14	30
16	18

Perfect score: 100   My score: _____

180

Answers
Math - Grade 7
(Answers for Pre-Tests and Tests are given on pages 191–192.)

Page 3

	a	b	c	d	e
1.	39	469	5459	18373	42546
2.	60	80	683	895	76696
3.	69	885	3978	44556	186979
4.	72	916	8785	66594	56293
5.	218	1767	9287	77598	86810
6.	226	2101	23836	86777	245948

Page 4

1. 449   3. 7473   5. 24338   7. 382002
2. 496   4. 27940   6. 71764

Page 5

	a	b	c	d	e
1.	53	433	598	4759	53471
2.	59	548	588	1546	38038
3.	12	251	509	3149	41381
4.	57	589	4111	2521	53517
5.	432	728	6539	4801	56262
6.	667	216	4681	5138	50469

Page 6

1. 126   3. 297   5. 2025   7. 11775
2. 186   4. 418   6. 2570

Page 7

	a	b	c
1.	121	2747	62541
2.	648	3518	86646
3.	254	7993	55209
4.	910	4468	130236
5.	179	11159	15798

Page 8

1. 870   3. 2173   5. 22277   7. 112554
2. 13   4. 794   6. 4034

Page 9

	a	b	c	d	e
1.	49	599	5753	25788	99975
2.	52	622	8331	62411	44511
3.	91	767	7884	88977	47583
4.	46	747	3528	53707	26567
5.	159	918	7776	78981	98922
6.	38	338	2174	62461	43496
7.	279	1986	17880	85483	203293
8.	43	217	1758	48729	48669

Page 10

1. 13   3. 1467   5. 263   7. 94710
2. 148   4. 2211   6. 49757

Page 11

	a	b	c	d
1.	306	684	2445	9363
2.	1425	7176	37968	50556
3.	6816	45368	297500	665199
4.	174870	274050	355950	2901465

Page 12

1. 325   3. 1024   5. 4000   7. 16875
2. 819   4. 7200   6. 52800   8. 224640

Page 13

	a	b	c	d
1.	50	60	50	50
2.	380	430	550	250

Page 13 (Continued)

3.	400	700	500	1000
4.	8700	3600	3300	2100
5.	5000	8000	7000	6000
6.	6000	8000	4000	10000

Page 14

	a			b	
1.	5952	6000	5544	5400
2.	1820	2000	2405	2800
3.	48804	50000	65682	64000
4.	121788	120000	199888	200000

	c	
1.	7371	7200
2.	3375	4000
3.	26104	25000
4.	184525	180000

Page 15

	a	b	c	d	e
1.	9 r6	16	12 r1	19 r2	12 r3
2.	27	156 r4	135	61 r2	168
3.	703	2541 r2	919 r1	855	331 r4

Page 16

1. 11 ; 4   3. 24   5. 95 ; 5
2. 28   4. 89   6. 73

Page 17

	a	b	c	d	e
1.	50	14 r8	10 r49	18 r11	11 r26
2.	6	8 r5	13 r30	12 r33	20 r27
3.	7 r69	15 r10	12 r63	4 r94	11 r20

Page 18

1. 12 ; 5   3. 25   5. 31 ; 20
2. 65   4. 42 ; 8   6. 18

Page 19

	a	b	c	d
1.	161 r8	412	39 r9	72
2.	1437	1069 r4	112 r48	201 r1
3.	5747	12071 r6	289 r27	2030

Page 20

1. 1074   3. 1554 ; 6   5. 144 ; 9
2. 2073   4. 44   6. 225

Page 21

	a	b	c
1.	64 r2	7 r47	56
2.	267 r5	68 r2	310 r6
3.	6865 r1	942	1283 r16

Page 22

1. 81   3. 1212   5. 23
2. 92   4. 12   6. 7 ; 71

Page 23

	a	b	c	d
1.	576	3800	3370	29250
2.	3591	22402	49875	439056
3.	127185	67983	1650384	775296
4.	8 r2	12 r72	57	80 r12
5.	136 r18	898 r2	922 r16	1858 r8

Answers Grade 7

Page 45

	a	b	c	d
1.	$\frac{13}{24}$	$\frac{11}{12}$	$\frac{17}{30}$	$1\frac{5}{18}$
2.	$3\frac{1}{20}$	$5\frac{19}{30}$	$10\frac{29}{36}$	$13\frac{8}{15}$
3.	$\frac{1}{12}$	$\frac{1}{18}$	$\frac{19}{48}$	$\frac{4}{75}$
4.	$1\frac{7}{12}$	$3\frac{1}{30}$	$4\frac{1}{40}$	$6\frac{13}{30}$

Page 46

1. $1\frac{7}{12}$ 3. $1\frac{13}{20}$ 5. red ; $\frac{1}{8}$ 7. $\frac{7}{12}$
2. Wednesday ; $\frac{1}{15}$ 4. $2\frac{5}{16}$ 6. Brenda ; $\frac{49}{60}$

Page 47

	a	b	c	d
1.	$6\frac{1}{8}$	$8\frac{7}{10}$	$2\frac{1}{2}$	$1\frac{3}{7}$
2.	$1\frac{3}{4}$	$2\frac{13}{15}$	$5\frac{1}{8}$	$\frac{5}{6}$
3.	$1\frac{1}{4}$	$1\frac{5}{6}$	$4\frac{1}{6}$	$2\frac{7}{24}$
4.	$4\frac{5}{6}$	$\frac{13}{30}$	$2\frac{35}{36}$	$4\frac{19}{20}$

Page 48

1. $1\frac{1}{4}$ 3. $2\frac{7}{20}$ 5. $1\frac{1}{2}$
2. $\frac{7}{30}$ 4. $1\frac{7}{40}$

Page 51

	a	b	c
1.	.9	3.9	.7
2.	.17	3.53	.07
3.	.259	4.357	.007
4.	5.5	12.40	4.14
5.	11.1	39.028	25.03
6.	$\frac{9}{10}$	$\frac{3}{10}$	$4\frac{1}{10}$
7.	$\frac{9}{10}$	$\frac{3}{10}$	$4\frac{1}{10}$
8.	$\frac{19}{100}$	$\frac{7}{100}$	$5\frac{3}{100}$
9.	$\frac{419}{1000}$	$\frac{11}{1000}$	$3\frac{333}{1000}$
10.	$13\frac{3}{10}$	$20\frac{27}{100}$	$20\frac{27}{1000}$
	$9\frac{3}{100}$	$100\frac{1}{10}$	$4\frac{567}{1000}$

Page 52

	a	b	c
1.	.2	.75	.018
2.	3.8	5.70	1.068
3.	5.5	4.48	9.615
4.	12.5	3.45	4.070
5.	$\frac{7}{10}$	$\frac{2}{5}$	$6\frac{1}{2}$
6.	$\frac{21}{100}$	$\frac{1}{50}$	$1\frac{3}{4}$
7.	$\frac{213}{1000}$	$\frac{1}{20}$	$5\frac{111}{200}$
8.	$\frac{3}{1000}$	$6\frac{9}{100}$	$10\frac{27}{250}$

Page 53

	a	b	c	d	e
1.	.8	1.1	7.5	13.4	13.2
2.	.86	.43	$.85	$7.39	$44.13
3.	.195	.419	.605	10.116	127.343
4.	1.75	.687	8.035	22.264	97.978
5.	.908	.667	24.95	3.269	44.833
6.	1.166	2.116	14.047	18.545	52.338

Page 54

1. 1.4 3. 57917.7 5. 360.8 7. 364.408
2. .051 4. 6.52 6. 3.608 8. Mable

Page 55

	a	b	c	d	e
1.	.6	.29	.093	2.049	38.531
2.	.594	1.39	$.58	1.379	$296.42
3.	.15	.136	3.86	8.293	528.42
4.	.086	.97	30.95	12.69	32.078
5.	4.4	6.89	6.583	41.784	63.725
6.	.216	6.762	3.327	34.192	35.575
7.	.082	1.837	5.942	6.712	2.074

Page 56

1. 2.9 3. 5.75 5. 2.86 7. 29.1
2. .68 4. .405 6. .017 8. 3.6

Page 57

1. 16.987 5. 18.798 9. 43.403
2. 3.88 6. 15.165 10. 3.82
3. 15.958 7. 25.902 11. 6.597
4. 1.708 8. 3.313 12. 9.93

Page 58

1. 1.418 3. .576 5. 3.461
2. 1.467 4. .603 6. more ; .581

Page 59

	a	b	c	d	e
1.	1.38	.80	6.4	1.620	301.7
2.	9.6	33.6	1.02	.6606	1.296
3.	.0092	.0085	.075	.1418	.03452
4.	.00252	.0304	.00304	.2048	.02418
5.	9.48	1.428	.6435	1.272	.3095
6.	1.263	4.904	.2772	.0418	1.248
7.	16.20	20.61	.3762	253.6	2.468

Page 60

	a	b	c	d	e
1.	.925	15.998	822.64	52.272	1.35141
2.	12.04	150.66	1.5662	1.31016	.35392
3.	.0496	1.7544	5.4730	4.3428	1.26072
4.	77.28	1.3338	13.5880	21.420	485.611
5.	138.75				
6.	13.875				
7.	3.276				

Page 61

	a	b	c
1.	83	830	8300
2.	.83	8.3	83
3.	754	7540	75400
4.	60.3	603	6030
5.	.345	3.45	34.5
6.	1002.5	10025	100250
7.	29.064	290.64	2906.4

Page 62

1. .003 3. 179.31 5. 4.5 ; 45 7. 55.40 ; 554
2. 7.638 4. 211.68 6. 5107.5

Page 63

	a	b	c	d
1.	.16	1.8	12.4	.318
2.	.09	.008	.023	2.027
3.	.004	.004	.008	7.68
4.	.018	.017	.028	.0014

Page 64

1. .042 3. 1.7 5. .37 7. .48
2. 1.44 4. .012 6. .94

183

Page 65

	a	b	c	d
1.	12	.94	3.6	.034
2.	180	1400	11500	7000
3.	20	120	15000	23

Page 66

1. 2.3 ; .6   3. .4   5. No ; 1.73   7. .02
2. 4 ; .24   4. 2.5   6. 50

Page 67

	a	b	c	d
1.	1070	13	.025	15
2.	27	1.3	560	350
3.	.9	.18	430	10
4.	340	120	6.3	1500

Page 68

1. 16   3. 5.26   5. 4.90   7. 11.2
2. 3   4. 1.2   6. 32   8. 100

Page 69

1. 136   5. 743.49   9. 6.3   13. 410
2. 2.32   6. 427.653   10. .0224   14. .54119
3. 34.98   7. .694   11. 127
4. 2.95   8. 144.6   12. 17.7241

Page 70

1. .706   4. 12.074   7. 8.56
2. 17.85   5. .13   8. 1.10
3. 175   6. 3.385   9. Ruth ; .002

Page 73

2. 9 to 2 $\frac{9}{2}$ 7. 6 to 2 $\frac{6}{2}$
3. 3 to 6 $\frac{3}{6}$ 8. 6 to 30 $\frac{6}{30}$
4. 8 to 9 $\frac{8}{9}$ 9. 3 to 1 $\frac{3}{1}$
5. 3 to 60 $\frac{3}{60}$ 10. 4 to 3 $\frac{4}{3}$
6. 15 to 5 $\frac{15}{5}$

Page 74

	a	b		a	b
1.	$\frac{2}{3}, \frac{6}{9}$	$\frac{7}{8}, \frac{13}{16}$	6.	$\frac{5}{6}, \frac{9}{12}$	$\frac{6}{16}, \frac{3}{8}$
2.	$\frac{5}{16}, \frac{11}{32}$	$\frac{4}{5}, \frac{8}{10}$	7.	$\frac{15}{20}, \frac{9}{10}$	$\frac{15}{24}, \frac{5}{8}$
3.	$\frac{5}{8}, \frac{10}{16}$	$\frac{3}{4}, \frac{16}{20}$	8.	$\frac{10}{12}, \frac{4}{5}$	$\frac{12}{8}, \frac{3}{4}$
4.	$\frac{1}{6}, \frac{2}{12}$	$\frac{3}{7}, \frac{9}{14}$	9.	$\frac{7}{25}, \frac{28}{100}$	$\frac{9}{20}, \frac{48}{100}$
5.	$\frac{3}{9}, \frac{1}{3}$	$\frac{3}{2}, \frac{18}{12}$	10.	$\frac{5}{16}, \frac{4}{12}$	$\frac{7}{8}, \frac{21}{24}$

Page 75

	a	b	c		a	b	c
1.	n=16	n=48	n=36	3.	n=45	n=3	n=4
2.	n=3	n=8	n=100	4.	n=8	n=3	n=70

Page 76

1. 24   3. 300   5. 65   7. 5
2. 2   4. 175   6. 147

Page 77

	a	b	c		a	b	c
1.	n=16	n=21	n=25	3.	n=6	n=16	n=6
2.	n=7	n=75	n=12	4.	n=72	n=875	n=49

Page 78

1. 600   3. 6
2. 2850   4. 300

Page 79

1. 4   3. .98   5. 2.97
2. .60   4. 2

Page 80

1. 14   3. 300   5. 1800   7. 9
2. 2   4. 70   6. 44

Page 81

	a	b	c		a	b	c
1.	n=1	n=4	n=5	5.	n=80	n=33	n=12
2.	n=24	n=6	n=13	6.	n=35	n=15	n=32
3.	n=6	n=12	n=26	7.	n=6	n=24	n=60
4.	n=96	n=20	n=7				

Page 82

1. 135   3. 375   5. 24   7. 27
2. 168   4. 16   6. 21   8. 20

Page 85

	a	b		a	b		a	b
1.	$\frac{1}{100}$	.01	6.	$\frac{143}{100}$	1.43	11.	$\frac{233}{100}$	2.33
2.	$\frac{21}{100}$	.21	7.	$\frac{9}{100}$	.09	12.	$\frac{83}{100}$	.83
3.	$\frac{129}{100}$	1.29	8.	$\frac{51}{100}$	.51	13.	$\frac{357}{100}$	3.57
4.	$\frac{3}{100}$	.03	9.	$\frac{169}{100}$	1.69	14.	$\frac{99}{100}$	.99
5.	$\frac{39}{100}$	.39	10.	$\frac{69}{100}$	.69			

Page 86

	a	b		a	b
1.	50	25	5.	30	$87\frac{1}{2}$
2.	10	$12\frac{1}{2}$	6.	80	24
3.	20	$37\frac{1}{2}$	7.	$33\frac{1}{3}$	70
4.	40	35	8.	$66\frac{2}{3}$	98

Page 87

	a	b		a	b
1.	$\frac{1}{4}$	$\frac{1}{2}$	7.	$\frac{3}{25}$	$\frac{19}{20}$
2.	$\frac{3}{10}$	$\frac{3}{4}$	8.	1	$1\frac{1}{2}$
3.	$\frac{2}{25}$	$\frac{4}{5}$	9.	$1\frac{3}{4}$	$1\frac{4}{5}$
4.	$\frac{3}{5}$	$\frac{1}{10}$	10.	$\frac{39}{50}$	$\frac{17}{20}$
5.	$\frac{1}{20}$	$\frac{2}{5}$	11.	$\frac{16}{25}$	$1\frac{9}{10}$
6.	$\frac{1}{5}$	$\frac{14}{25}$	12.	$3\frac{1}{2}$	$\frac{1}{50}$

Page 88

	a	b		a	b
1.	60	$62\frac{1}{2}$	7.	$\frac{39}{50}$	$1\frac{2}{5}$
2.	75	90	8.	$\frac{3}{50}$	$\frac{31}{50}$
3.	16	34	9.	$\frac{9}{20}$	$\frac{33}{50}$
4.	$66\frac{2}{3}$	55	10.	$\frac{41}{50}$	$\frac{13}{50}$
5.	$\frac{3}{4}$	$\frac{9}{10}$	11..	$1\frac{1}{5}$	$\frac{9}{50}$
6.	$2\frac{1}{2}$	$\frac{7}{25}$	12.	$\frac{1}{25}$	$3\frac{1}{4}$

Page 89

	a	b	c
1.	70	90	50
2.	130	260	190
3.	3	84	35
4.	106	187	245
5.	1.5	22.5	37.5
6.	103.2	212.5	147.7
7.	8	87.5	375
8.	124	130	200.1
9.	10	39.6	17
10.	129.6	239	230

Page 90

	a	b	c
1.	.85	.12	.35
2.	.04	.06	.07
3.	.25	.75	.33
4.	.994	.5625	.1875

Page 90 (continued)

5.	1	.625	.875
6.	.164	.9375	2.50
7.	.096	.0625	4.75
8.	.54	1.25	.195

Page 91

	a	b	c
1.	.085	.0775	.064
2.	.0525	.125	.081
3.	.1875	.0625	.078
4.	.375	.0975	.052
5.	.0825	.875	.096
6.	.6875	.073	.075
7.	.084	.1625	.097
8.	.065	.9375	.082
9.	.374	.035	.3125
10.	.839	.426	.2175
11.	.0214	.1225	.014

Page 92

1.	50%	.074	9.	.7%	.055
2.	170%	.168	10.	141.4%	.0725
3.	280%	1.375	11.	8.75%	.0475
4.	9%	.0749	12.	.95%	.9375
5.	67%	.8667	13.	106.55%	.625
6.	115%	2.0986	14.	.09%	.354
7.	7.5%	.07125	15.	31.25%	.093
8.	37.5%	.0375			

Page 93

	a	b
1.	$\frac{1}{20}$	.08
2.	10%	9%
3.	$\frac{3}{25}$	.12
4.	$12\frac{1}{2}\%$	27%
5.	$\frac{3}{20}$	.075
6.	20%	16.8%
7.	$\frac{1}{4}$	1.25
8.	75%	135%
9.	$\frac{37}{100}$	.065
10.	40%	27.48%
11.	$\frac{1}{2}$	2.065
12.	60%	137.5%
13.	$\frac{31}{50}$	.0825
14.	$87\frac{1}{2}\%$	6.75%
15.	$\frac{83}{100}$	.0775

Page 94

1. 25 3. 25 5. $\frac{3}{4}$
2. .065 4. $\frac{1}{4}$

Page 97

	a	b		a	b
1.	40	30	6.	7.65	49
2.	9	9.6	7.	27.645	65.7
3.	7.5	7.2	8.	4.5	4.5
4.	19.2	80	9.	10	7
5.	9.6	48			

Page 98

1. 12 3. 441 5. 60 7. 108
2. 240 4. 56 6. 1440

Page 99

	a	b		a	b
1.	20	37.5	6.	12.5	125
2.	25	50	7.	45.5	20
3.	100	90	8.	200	75
4.	8.4	75	9.	1.1	150
5.	125	175	10.	60	81.25

Page 100

1. 80 3. 20 5. 25 7. $62\frac{1}{2}$
2. 40 4. $82\frac{1}{2}$ 6. 5

Page 101

	a	b		a	b
1.	60	2	6.	10	.4
2.	18	200	7.	750	5.12
3.	160	740	8.	240	60
4.	5	83	9.	50	80
5.	3.2	7	10.	100	30

Page 102

1. 15 3. 210 5. 50 7. 15.60
2. 500 4. $1\frac{1}{4}$ 6. 325

Page 103

	a	b		a	b
1.	180	507	7.	6	9
2.	80	240	8.	20	25
3.	39.2	2.4	9.	140	60
4.	6.27	612	10.	100	780
5.	25	75	11.	1	3.75
6.	250	100	12.	.4	30.6

Page 104

1. 2400 3. 35 5. 13,000
2. 2472.50 4. 15,200

Page 105

	a	b		a	b
1.	27	200	7.	45	90
2.	75	3.15	8.	20	3.3
3.	40	120	9.	1480	50
4.	22.75	900	10.	63.7	968
5.	43.75	16	11.	150	135
6.	50	95	12.	140	37.5

Page 106

1. 39 3. 36 5. 87.5 7. 60
2. 35 4. 30 6. 240

Page 109

1. $10 5. $2.10 9. $2.14
2. $12 6. $3.12 10. $25.67
3. $20 7. $33.20 11. $5.16
4. $4.50 8. $55.50 12. $66.01

Page 110

1. 40.50 3. 1.25 5. 120 7. Marge ; 1
2. 3.90 4. 94.50 6. 50.49

Page 111

1.	$90	6.	$120	$620
2.	$112	7.	$252	$852
3.	$475	8.	$161.50	$1011.50
4.	$273.80	9.	$1275	$3775
5.	$753.75	10.	$1167	$5057

Page 112

1. 135	3. 100	5. 1361.25	7. 5386.50
2. 600 ; 3100	4. 540	6. 1402.50	

Page 113

1. $120	6. $105	$805
2. $90	7. $5.10	$90.10
3. $73.50	8. $71.25	$1071.25
4. $589.05	9. $918.75	$2793.75
5. 618.75	10. $341.25	$1741.25

Page 114

1. 144	3. 26	5. 3956.25
2. 40.05	4. 30 ; 780	6. Marvin ; .90

Page 115

1. $2	5. $23.10	8. $54
2. $4	6. $87.75	9. $12
3. $3	7. $10	10. $118.75
4. $15		

Page 116

1. 15	3. 114 ; 2014	5. 900
2. 3.50	4. 4.50	6. 7095 ; 93095

Page 119

	a	b	c
1.	9000	1200	6000
2.	30	5300	40
3.	92	990	10000
4.	10540	1280	1200
5.	1	100000	450
6.	200000	187	10

7. less   8. 1500

Page 120

	a	b	c
1.	13.0	.028	.345
2.	600	.100	9250
3.	.009	2050	1.500
4.	.206	24850	4.8
5.	.999	9890	1
6.	.050	1.526	625

7. 23.8   8. Alex

Page 121

	a	b
1.	8000	17000
2.	.200	6000
3.	2.000	3.440
4.	10000	.045
5.	100	940
6.	.2405	.236
7.	3.500	4500
8.	.001	30800

9. 1000000   10. 1020

Page 122

	a	b
1.	3000	10000
2.	.485	.250
3.	3500	4.600
4.	45800	2.5009

5. 2.5 ; 2500   6. 9.5 ; 9500   7. 10

Page 123

	a	b
1.	6000	8500
2.	.675	5.0505

3. 1.6   5. 7500   7. 350000
4. 1.15   6. 1.260

Page 124

1. 3.25 ; 3250	3. 200	5. .93	7. 600
2. 3500	4. 1.450	6. .80	8. 3500

Page 127

	a	b
2.	SR or RS	$\overleftrightarrow{SR}$ or $\overleftrightarrow{RS}$
3.	YX or XY	$\overleftrightarrow{YX}$ or $\overleftrightarrow{XY}$
4.	PQ or QP	$\overleftrightarrow{PQ}$ or $\overleftrightarrow{QP}$

	a	b	c
6.	JK or KJ	$\overline{JK}$ or $\overline{KJ}$	J and K
7.	RS or SR	$\overline{RS}$ or $\overline{SR}$	R and S
8.	BG or GB	$\overline{BG}$ or $\overline{GB}$	B and G

Page 128

	a	b	c
2.	QP	$\overrightarrow{QP}$	Q
3.	XY	$\overrightarrow{XY}$	X
4.	KJ	$\overrightarrow{KJ}$	K
6.	JKL or LKJ KJ and KL	∠JKL or ∠LKJ	
7.	PQR or RQP QP and QR	∠PQR or ∠RQP	

Page 129

2. intersecting ; parallel   4. 0   6. 1
3. parallel ; intersecting   5. No   7. Yes

Page 130

	a	b	c	d		a	b	c	d
1.		R		R	3.	P			P
2.	R		R						

Page 131

	a	b	c
1.	obtuse	right	acute
2.	acute	obtuse	right
3.	acute	right	obtuse

Page 132

	a	b	c
1.	right	obtuse	acute
2.	acute	right	obtuse
3.	obtuse	acute	right

Page 133

	a	b	c
1.	scalene	isosceles or equilateral	isosceles
2.	isosceles or equilateral	isosceles	scalene
3.	scalene	isosceles or equilateral	isosceles

Page 134

	a	b
1.	figures EFGH and PQRS	figure PQRS
2.	Yes	No
3.	figures JKLM and PQRS	figure PQRS
4.	Yes	No

Page 137

	a	b	c
1.	20	12	12
2.	35	30	168

3. 26   6. 11
4. 16   7. 28
5. 20

Page 138

1. 542   3. 14   5. short ; 10
2. 446   4. $10\frac{1}{4}$ 6. triangular board ; .5

Page 139

	a	*b*	*c*
1.	66	44	$31\frac{3}{7}$
2.	$16\frac{1}{2}$	$36\frac{2}{3}$	22

3. 4 ; 25.12
4. 6 ; 18.84
5. $3\frac{1}{2}$; 21.98
6. 17 ; 53.38
7. 4.7 ; 29.516
8. 14.4 ; 45.216

Page 140

1. 14 ; 44
2. 14 ; 88
3. $103\frac{5}{7}$
4. 9 ; 56.52
5. 7.85
6. square ; 2.58

Page 141

1. 7 ; $3\frac{1}{2}$
2. 35 ; $17\frac{1}{2}$
3. 28 ; 14
4. $3\frac{1}{2}$; $1\frac{3}{4}$
5. $\frac{7}{11}$; $\frac{7}{22}$
6. $10\frac{1}{2}$; $5\frac{1}{4}$
7. 2 ; 1
8. 10 ; 5
9. 3 ; $1\frac{1}{2}$
10. 7 ; $3\frac{1}{2}$
11. 1 ; $\frac{1}{2}$
12. 4 ; 2

Page 142

1. 35 ; $17\frac{1}{2}$
2. $1\frac{5}{44}$; $2\frac{5}{22}$
3. 1 ; $\frac{1}{2}$
4. 1.5 ; 3
5. .5 ; .25
6. 2 ; 1

Page 143

	a	*b*	*c*
1.	32	64	153

2. 54
3. 56.25
4. $15\frac{3}{4}$
5. $5\frac{1}{16}$
6. 16.25
7. 34.03

Page 144

1. $93\frac{1}{2}$
2. $30\frac{1}{4}$
3. 558
4. Yes
5. square ; 1
6. 8.4 ; 18.9

Page 145

	a	*b*	*c*
1.	27	6	30

2. 36
3. 30
4. 7
5. 30.75
6. $7\frac{7}{16}$
7. 86

Page 146

1. 27
2. $6\frac{3}{4}$
3. 1134
4. $24\frac{1}{2}$
5. 7.2 ; 12.8
6. 2.56

Page 147

	a	*b*	*c*
1.	28.26	254.34	113.04
2.	50.24	12.56	78.5

	d	*r*	*A*
3.	42		1386
4.		7	154
5.	32		$804\frac{4}{7}$
6.		35	3850
7.	56		2464

Page 148

1. $12\frac{4}{7}$
2. 5544
3. 1386
4. 22176
5. 452.16
6. 28.26 ; 36
7. B ; 15.7

Page 149

	a	*b*		*a*	*b*
1.	64	12	3.	16	40
2.	$28\frac{1}{2}$	30	4.	$15\frac{3}{4}$	226.08

Page 150

1. 1728
2. 52
3. 16
4. 31
5. 4904
6. 352

Page 153

	a	*b*	*c*
1.	210	27	49.5
2.	144		

3. 180
4. 40.96
5. $35\frac{7}{16}$
6. $8\frac{3}{4}$

Page 154

1. $157\frac{1}{2}$
2. 1296
3. 10368
4. 1728
5. 27
6. Jeff's ; 4
7. 60

Page 155

	a	*b*	*c*
1.	60	105	21
2.	135	$\frac{1}{4}$	576

Page 156

1. 12
2. 6
3. 1.35
4. 48
5. 3.6 ; 18 ; 21.6

Page 157

	a	*b*	*c*
1.	$6\frac{2}{7}$	$113\frac{1}{7}$	$1527\frac{3}{7}$

	d	*r*	*V*
2.	6		169.56
3.	14		615.44
4.		5	235.5
5.	10		157
6.		9	763.02

Page 158

1. 226.08
2. 113.04
3. 150.72
4. 452.16
5. 803.84
6. 157
7. flour ; 549.5

Page 161

1. Answers may vary. One difference is that the bars run in different directions.
2. 20
3. Judy
4. Mona
5. Judy
6. 10

Page 162

1.

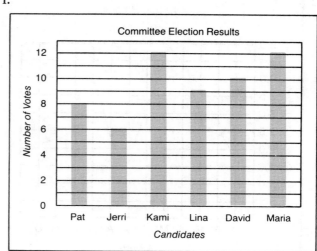

187

Page 162 (continued)

2.

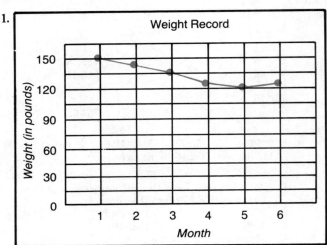

Cost of Parts

Part A
Part B
Part C
Part D

.00 .50 1.00 1.50 2.00 2.50 3.00

Cost (in dollars)

Page 163

1. December
2. June, July, August
3. February
4. 5 inches
5. June, March
6. February, April
7. decrease

Page 164

1.

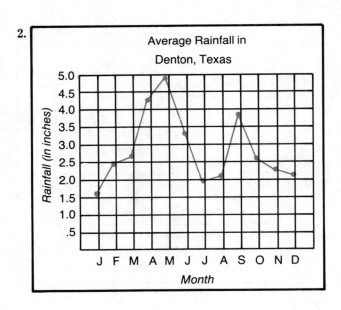

Weight Record

150
120
90
60
30
0

Weight (in pounds)

1   2   3   4   5   6

Month

2.

Average Rainfall in
Denton, Texas

5.0
4.5
4.0
3.5
3.0
2.5
2.0
1.5
1.0
.5

Rainfall (in inches)

J F M A M J J A S O N D

Month

Page 165

1. Answers may vary. Graph B used smaller intervals between months and greater intervals between amounts of sales, therefore making sales look more impressive.
2. 30
3. 20
4. 50
5. decreasing
6. By reversing the months, if a person does not look carefully, it appears that sales are increasing.

Page 166

1. 400
2. 300
3. 200
4. decreasing
5. increasing

6.

25
20
15
10
5
0

Number Produced

1 2 3 4 5 6 7 8 9 10

Week

Key:
—— Workers Absent
—— Number Produced

Page 167

1. $\frac{1}{4}$
2. $\frac{1}{4}$
3. $\frac{3}{4}$
4. $\frac{1}{4}$
5. $\frac{1}{4}$
6. $\frac{1}{2}$
7. $\frac{1}{6}$
8. $\frac{1}{3}$
9. $\frac{1}{2}$
10. $\frac{5}{6}$

Page 168

1. $\frac{1}{2}$
2. $\frac{1}{6}$
3. $\frac{1}{4}$
4. $\frac{1}{3}$
5. $\frac{2}{3}$
6. $\frac{3}{8}$
7. $\frac{1}{8}$
8. $\frac{7}{8}$

Page 169

1. 8
2. $\frac{1}{2}$
3. 0
4. $\frac{1}{8}$
5. $\frac{7}{8}$
6. 1
7. $\frac{3}{8}$
8. 1

Page 170

1. $\frac{1}{4}$
2. $\frac{1}{5}$
3. 0
4. $\frac{1}{2}$
5. 1
6. 0
7. 1

Page 171

1. 9
2. $\frac{1}{9}$
3. 8
4. $\frac{1}{8}$
5. $\frac{1}{4}$
6. $\frac{1}{2}$

Page 172

1. $\frac{1}{36}$
2. $\frac{1}{36}$
3. $\frac{1}{18}$
4. $\frac{1}{12}$
5. 0
6. 1

Page 173

1. $\frac{1}{1000}$
2. $\frac{1}{200}$
3. $\frac{1}{40}$
4. $\frac{939}{1000}$
5. $\frac{1}{2}$
6. $\frac{3}{10}$
7. 0

Page vii
1. 23; 8; multiply; 184; yes
2. add; $5.68; yes

Page viii
1. 2,400; 60; divide; 40; yes
2. Machine B—50 units/hour and Machine A—3 times that many units; multiply; 150; yes
3. add; $10,300; yes

Page ix
1. Units made—234, 199, 232, 219; add; 884
2. 15 days of vacation and 9 used; subtract; 6
3. 232 units in 29 hours; divide; 8

Page x
1. 50 ; 5   2. 15 ; 24 ; 4th

Page xi
1. 55   **2.** 676   **3.** 21   **4.** second

Page xii
1. 10   2. 10
3. Answers will vary. Should include:
   Fewer misunderstandings
   Have it in writing for protection

Page xiii
1. 4 ; 3   3. 3 ; 12
2. 5 ; 7   4. 13 ; 3
5. Answers will vary. The main problem is not knowing how much each person started with.

Page xiv
1. 8   4. 6
2. fourth   5. 3
3. Jane   6. 14 , 8 ; 10 , 14 ; 12, 8 ; 16, 22

Page 1

	a	b	c	d	e
1.	84	792	2028	20506	92607
2.	34	471	446	4333	36125
3.	134	1570	4984	11800	
4.	18272	283122	50244	328130	
5.	24 r1	98 r1	8 r29	16 r3	
6.	524 r2	8051	127 r65	236 r24	

Page 2
1. 24   2. 54
3. Brea and Joji ; 21

Page 25

	a	b	c	d	e
1.	79	1097	7015	10745	97989
2.	25	68	2385	54206	54947
3.	342	2085	21024	178038	2383094
4.	9 r2	158 r7	22 r3	901	2150 r2

Page 26

	a	b	c	d
1.	$\frac{8}{15}$	$\frac{7}{10}$	$\frac{3}{4}$	$\frac{1}{2}$
2.	$28\frac{1}{2}$	4	$5\frac{13}{24}$	$8\frac{2}{5}$
3.	$1\frac{1}{3}$	$\frac{3}{4}$	8	$3\frac{4}{7}$
4.	$\frac{7}{10}$	$\frac{3}{4}$	$7\frac{29}{30}$	$11\frac{7}{40}$
5.	$\frac{2}{9}$	$\frac{2}{15}$	$3\frac{1}{2}$	$2\frac{23}{30}$

Page 49

	a	b	c	d
1.	$\frac{4}{9}$	$\frac{2}{3}$	$\frac{4}{15}$	$\frac{1}{4}$

Page 49 (Continued)

2.	90	6	$6\frac{2}{3}$	$38\frac{1}{2}$
3.	$\frac{24}{25}$	$2\frac{11}{12}$	$10\frac{2}{3}$	4
4.	$\frac{4}{5}$	$1\frac{4}{9}$	$8\frac{1}{12}$	$3\frac{2}{15}$
5.	$\frac{3}{5}$	$\frac{27}{56}$	$4\frac{3}{4}$	$1\frac{7}{10}$

Page 50

	a	b	c	d	e
1.	9.5	2.109	48.49	4.949	24.831
2.	.49	38.91	5.133	40.413	31.76
3.	2.837	19.303	91.819	43.36	66.661
4.	3.48	1.46	.00392	.924	
5.	1.9	.0038	315	5.4	
6.	22.1292	3.02736	.37	370	

Page 71

	a	b	c	d
1.	2.30	21.733	114.11	17.872
2.	17.729	3.37	2.42	25.711
3.	8.65	.204	.0648	1.12
4.	1.4	.0142	.313	4200
5.	1.43242	609.444	4.7	.41

Page 72

	a	b
2.	2 to 10	$\frac{2}{10}$
3.	8 to 2	$\frac{8}{2}$
4.	3 to 5	$\frac{3}{5}$

5. a) $\frac{5}{8}, \frac{15}{24}$ b) $\frac{7}{12}, \frac{11}{18}$
6. a) $\frac{3}{4}, \frac{16}{20}$ b) $\frac{2}{3}, \frac{32}{48}$
7. a) $\frac{4}{5}, \frac{24}{30}$ b) $\frac{5}{6}, \frac{20}{24}$

	a	b
8.	$n = 27$	$n = 16$
9.	$n = 7$	$n = 6$
10.	$n = 60$	$n = 5$

Page 83

	a	b
1.	28 to 3	$\frac{28}{3}$
2.	5 to 2	$\frac{5}{2}$
3.	17 to 200	$\frac{17}{200}$
4.	4 to 7	$\frac{4}{7}$

	a	b	c
5.	$n = 32$	$n = 14$	$n = 4$
6.	$n = 12$	$n = 32$	$n = 750$
7.	$n = 27$	$n = 20$	$n = 72$

8. 30   9. 316 ; 395

Page 84

	a	b		a	b
1.	70	75	8.	.0775	.0925
2.	90	14	9.	$\frac{27}{100}$	20
3.	130	$37\frac{1}{2}$	10.	$1\frac{17}{100}$	$2\frac{4}{5}$
4.	72.5	167.2	11.	$\frac{27}{50}$	$1\frac{3}{4}$
5.	.07	.69	12.	$\frac{3}{100}$	$\frac{1}{20}$
6.	.125	.0325			
7.	.1876	.07125			

Answers

Page 95

1. $\frac{9}{100}$ .09   3. $\frac{79}{100}$ .79
2. $\frac{47}{100}$ .47   4. $\frac{153}{100}$ 1.53

	a	b	c
5.	30	62.5	80
6.	120	32.5	31.75
7.	.6225	1.265	.0875
8.	$\frac{17}{20}$	$1\frac{3}{5}$	$\frac{3}{50}$

Page 96

	a	b
1.	32	70
2.	$11\frac{1}{2}$	$15\frac{3}{4}$
3.	$4\frac{1}{5}$	150
4.	59.2	67.2
5.	25	$62\frac{1}{2}$
6.	75	120
7.	50	16
8.	$31\frac{1}{4}$	20
9.	70	140
10.	140	300
11.	100	70
12.	.5	132

Page 107

	a	b			a	b
1.	80	50	6.		$56\frac{1}{4}$	300
2.	18	70	7.		26.68	25
3.	500	75	8.		20	135
4.	30	58.5	9.		100	2
5.	2	336	10.		62.56	$87\frac{1}{2}$

Page 108

1. $32 $432
2. $78 $728
3. $18 $138
4. $176 $976
5. $121.50 $571.50
6. $37.50   10. $14   14. $610.50
7. $81   11. $35.10   15. $174.90
8. $4.05   12. $13.50
9. $.80   13. 123.75

Page 117

1. $45 ; $545   9. $45.50
2. $44 ; $444   10. $115.50
3. $180 ; $780   11. $20
4. $180 ; $930   12. $57.50
5. $570 ; $3570   13. $136.40
6. $67.50   14. $186
7. $123.75   15. $896
8. $132

Page 118

	a	b
1.	90	1000
2.	2500	1000
3.	3.0	3.500
4.	2500	46870
5.	.49	.250
6.	4000	1700

Page 118 (continued)

	a	b
7.	.500	2.480
8.	10800	.800
9.	100000	.2605
10.	4.800	900

11. 2   12. 28   13. 904

Page 125

	a	b
1.	120	6000
2.	3000	2000
3.	8.0	.900
4.	6400	54920
5.	.5492	.505
6.	6000	1200
7.	700	.085
8.	1.200	3.500
9.	15900	.075
10.	250000	.875
11.	10.000	750

12. 969.6   13. 1500   14. 20

Page 126

	a	b	c	d
2.	$\overline{JL}$ or $\overline{LJ}$	$\overleftrightarrow{AB}$ or $\overleftrightarrow{BA}$	∠CDE or ∠EDC	$\overrightarrow{GF}$

	a	b	c
3.	isosceles	equilateral or isosceles	scalene
4.	R	R, S, X	X
5.	parallel	intersecting	intersecting

Page 135

	a	b	c
1.	$\overline{ME}$ or $\overline{EM}$	$\overrightarrow{DR}$	∠PTA or ∠ATP
2.	right	obtuse	acute

3. a   5. a, b, c, d
4. a, c   6. a, b

Page 136

	a	b	c		a	b	c
1.	30	20.8	176	4.	153.86	42	$56\frac{1}{4}$
2.	12	308	$16\frac{1}{4}$	5.	8.82	28.26	1050
3.	14	25.2	50.24				

Page 151

1. 24 ; 35   6. 19
2. 9 ; 4.5   7. 12.75
3. 10 ; $5\frac{1}{4}$ 8. 4 ; 25.12 ; 50.24
4. 20 ; 25   9. 200 ; 628 ; 31400
5. 28   10. 6 ; 3 ; 28.26

Page 152

	a	b	c
1.	36	64	$67\frac{1}{2}$
2.	500	36	1.6
3.	706.5	502.4	1356.48
4.	15	4.5	226.08

Page 159

1. 192   3. 405
2. 343   4. $38\frac{1}{2}$

	a	b	c
5.	60	768	37.5

	d	r	V
6.	24		3617.28
7.		8	3818.24
8.	20		3768
9.		10.5	4846.59

1. Pam 4. $120,000
2. Jan 5. $30,000
3. Ed
6. Answers may vary. One of the following should be included: No label on left side. Years are reversed.

7. $\frac{1}{6}$ 9. $\frac{1}{3}$ 11. $\frac{1}{2}$
8. $\frac{2}{3}$ 10. $\frac{1}{3}$

Page 174

1. Roi 6. $\frac{1}{6}$ 9. $\frac{1}{3}$
2. Mia 7. 0 10. $\frac{2}{3}$
3. 10 8. $\frac{1}{2}$
4. 3

5.

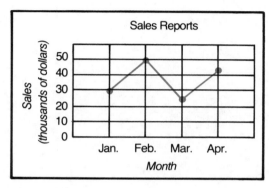

Sales Reports

Page 175

	a	b	c	d	e
1.	93350	1303	89	21489	1989
2.	30240	2943052	13 r4	206 r8	156 r46
3.	$47.56	1.7983	.368	$11.35	10.173
4.	.486	57.9568	.201	320	1.52
5.	$\frac{4}{7}$	$3\frac{1}{2}$	$\frac{1}{8}$	$\frac{3}{4}$	$2\frac{1}{6}$
6.	$1\frac{17}{24}$	$5\frac{17}{30}$	$\frac{2}{9}$	$1\frac{1}{8}$	$7\frac{17}{20}$

Page 176

	a	b	c	d
7.	24	40	10	2
8.	6	16	95	250

	a	b	
9.	$\frac{1}{10}$	.06	13. 20
10.	50%	28.6%	14. 33.67
11.	$\frac{17}{20}$	1.055	15. 25
12.	87.5%	8.5%	16. 9

Page 177

	a	b	c	d	e
1.	104320	1153	383	69147	22785
2.	5058144	1169	872 r25	$11.05	$17.90
3.	.3924	661.9712	.201	320	1.52
4.	$\frac{7}{15}$	7	$1\frac{2}{7}$	2	
5.	$1\frac{9}{35}$	$9\frac{1}{6}$	$\frac{7}{30}$	$2\frac{3}{8}$	
6.	2	27	90	6	
7.	125	80	12	9	

Page 178

	a	b	c
8.	$\frac{9}{50}$; .18	$\frac{3}{50}$; .06	
9.	25% ; .25	5% ; .05	
10.	7% ; $\frac{7}{100}$	110% ; $1\frac{1}{10}$	
11.	32	50%	2.75
12.	200	10%	6000
13.	27.5%	3.65	70
14.	90%	800	.45
15.	$64	$864	
16.	$210	$1210	
17.	$304.20	$1084.20	
18.	$33.75	$933.75	
19.	$75	$2075	
20.	190000	8	600
21.	8000	900	.075
22.	28000	.75	2000

Page 179

23. i 26. a 29. m 31. l
24. d 27. k 30. c 32. g
25. e 28. h

	a	b	c
33.	49	54	50.24
34.	35	36	25.12

Page 180

	a	b	c
35.	1485	105.6	84
36.	288	477.28	529.875

37. 125 38. E 39. $\frac{5}{8}$ 40. 1

Photo Credits

Tony Freeman, 104; Glencoe stock, 94; Stephen McBrady, 2; Cameron Mitchell, 48, 79; Courtesy MTI Corporation, 58; Alan Oddie/Photo Edit, 44